Endorsements

"*Success is Assured* is a practical guide to successful product development using 'Set-Based Design' principles and an 'Intelligent Fast Failure' approach. The authors rely on their extensive industry experience in their respective fields and as successful industry consultants helping to transform the way existing companies do business. An easy-to-read addition to the field with many new, useful conceptual models."

Charlie Camarda
Astronaut STS-114 Discovery (Return to Flight mission),
Former Director of Engineering, NASA Johnson Space Center

"In 1953, my dad led his development team at GM to design and build the iconic '55 Chevy in just 23 months—unheard of then and certainly not possible today. How did they succeed in that? Through focused learning and decision making leveraging strong engineering expertise. Over the last decade, we have strayed from that, focusing too much on tasks and schedules. This book shows us how to rebuild that capability through new analytical tools for decision-focused learning and the creation and reuse of visual knowledge. This is increasingly important as complexity is increasing dramatically and the need for effective and speedy product development is critical to success."

David Cole, Ph.D.
Chair, AutoHarvest; Chair Emeritus, Center for Auto Research;
Chair, Building America's Tomorrow; Automotive Hall of Fame Inductee

"I wish this book was already written when we started our Lean Journey almost 11 years ago. It is a breakthrough in the thinking process on how to make decisions that will drive your business results to 'Success is Assured'! From the very core definition of the 'True North' of Product Development to the Enabler of all Enablers (the Causal Map), the authors' experience provides everyone clarity on how to learn faster and innovate better while having a lot more fun! My team and I are very enthusiastic and we almost cannot hold ourselves in anticipation of the great achievements we'll be able to provide to our business partners."

Manoel de Queiroz Cordova Santos
Business Excellence Manager—Product Development, Embraer S.A.

"Transforming New Product Development for complex products is a difficult endeavor. As if the technical challenges generated by the large number of variables and interdependencies were not enough, the current leadership practices could drive the wrong behaviors, creating additional complications. What new behaviors should be embraced by leadership? How to tackle complicated technical interdependencies in a new and robust way? This book does an excellent job in covering both questions, using an approach that includes real and fictional stories, peppered with a lot of examples. Highly recommended for both leadership and technical experts embarking in their New Product Development transformation."

Ovidiu Contras
Lean Coach, Bombardier Aerospace

"This book is invaluable for any team or organization aiming to design complex systems successfully. The practical knowledge based approach for robust decision making explained in the book, with lucid examples, is exceptionally relevant. The case study story is something most teams and organizations dealing with the design and development of complex systems can easily correlate with."

Ramakrishnan Raman, ESEP
Assistant Sector Director—INCOSE Asia Oceania;
Principal Systems Engineer—Honeywell

"As for any company in charge of developing new products, the key challenge is to define the solution that provides the best trade-offs between customer's needs and the Company's goals and capabilities to provide maximum value to both. The traditional Point-Based Design quickly converges on a solution based on a single point in the design space—a process which inevitably creates rework, delays and cost overruns. This book explains in a very effective instructive style the Set-Based Design approach—a most powerful method to develop better solutions. The key principle is to focus upfront in building key knowledge to make decisions by exploring the trade-offs between critical parameters and mapping an area of feasibility. This is the first book that describes Set-Based Design in an easy to apply way. It provides a set of powerful tools and methods developed by the authors and shows how to apply them through many examples. It provides clarity in how to conduct the development process and build it step by step with different enablers to assure success! This book is indispensable for anyone willing to make their process of product development much more valuable."

Luc Delamotte
Senior Coach Lean Engineering, Lean Community Manager, Thales

"I have always liked Michael´s approach to this complex subject. He has always been curious to learn more and challenge every 'truth' he encountered on his way seeking 'Success is Assured'. In this book the authors capture the true essence of the matter, what it takes to assure the success of the product development project. A real 'how to' book which will help many product developers to shorten lead times and to innovate! Read the book and apply the knowledge and your success will be assured."

Peter Palmér
Chairman, Lean Product & Process Development Exchange (LPPDE);
Senior Manager, R&D Way Office, Scania

"I highly recommend *Success is Assured*! To my knowledge, it is the only book in print that provides a detailed methodology on the application of set-based design to complex system development using real-world examples."

Donny Blair
Senior Director of Engineering, L3 Technologies

"If you are still not convinced about the merits of Lean Product Development, read this book and you will be. Even if you are not an engineer."

Dr. Göran Gustafsson
Chalmers University of Technology

Success is Assured

Satisfy Your Customers On Time and On Budget by Optimizing Decisions Collaboratively Using Reusable Visual Models

By
Penny W. Cloft
Michael N. Kennedy
Brian M. Kennedy

Routledge
Taylor & Francis Group

A PRODUCTIVITY PRESS BOOK

On the cover: The U.S. Army testing the airplane Orville Wright delivered in July 1909 at Fort Meyer in Virginia. The U.S. Army offered to pay the Wright Brothers $25,000 if they could deliver an airplane that met their specifications, which included being able to fly 10 miles with a passenger at an average speed of 40 mph; and they offered an additional $2,500 for each full mile per hour over 40 mph. It succeeded in meeting all of the specs and averaged 42.5 mph, and thus they were paid $30,000 for that airplane.

First edition published in 2019

by Routledge/Productivity Press
711 Third Avenue New York, NY 10017, USA
2 Park Square, Milton Park, Abingdon, Oxon OX14 4RN, UK

© 2019 by Penny W. Cloft, Michael N. Kennedy and Brian M. Kennedy
Routledge/Productivity Press is an imprint of Taylor & Francis Group, an Informa business

No claim to original U.S. Government works

Printed on acid-free paper

International Standard Book Number-13: 978-1-138-61858-9 (Hardback)

Library of Congress Cataloging-in-Publication Data

Names: Cloft, Penny W., author. | Kennedy, Michael N., author. | Kennedy, Brian M., author.
Title: Success is assured : satisfy your customers on-time and on-budget by optimizing decisions collaboratively using reusable visual models / by Penny W. Cloft, Michael N. Kennedy, Brian M. Kennedy.
Description: 1 Edition. | Boca Raton, Fla. : CRC Press, [2019] | Includes bibliographical references and index.
Identifiers: LCCN 2018013125| ISBN 9781138618589 (hardback : alk. paper) | ISBN 9781315226767 (ebook)
Subjects: LCSH: Consumer satisfaction. | Success in business. | Decision-making.
Classification: LCC HF5415.335 .C56 2019 | DDC 658.8/343--dc23
LC record available at https://lccn.loc.gov/2018013125

Visit this book's website at
http://www.successisassured.com

and the Targeted Convergence website at
http://www.targetedconvergence.com

Targeted Convergence, Set-Based Thinking, and Success Assured are trademarks or registered trademarks of Targeted Convergence Corporation in the United States or other countries.

Visit the Taylor & Francis Web site at
http://www.taylorandfrancis.com

Contents

Foreword

My name is Ron Marsiglio. I was president and CEO of a Teledyne company, Benthos, which designs and builds undersea acoustic systems and TapTone brand industrial testing equipment. We were one of the first companies to adopt the practices discussed in this new book. We at Benthos proved that it is indeed possible to establish "Success is Assured" before beginning the detailed design of a new product. Our journey is well documented in the predecessor to this book, *Ready, Set, Dominate*.

The premise we followed was actually quite simple: to first understand your True Customer Interests, you must know what end user problem you are solving and those needs must be translated into easily measured engineering design criterion. Then you must identify all known design knowledge gaps, which are related to solving your customer's problem. Then you must relentlessly close those knowledge gaps before you start production intent product design. Success requires that. Hoping knowledge gaps will somehow be magically closed later in the process cannot be tolerated. You must banish all wishful thinking.

We began by learning the methodology and tools derived from the Toyota Product Development System: Look, Ask, Model, Discuss, Act (LAMDA) and K-Briefs (A3s) for problem-solving as well as understanding and capturing true customer interests. We also learned how to identify the causal interactions and trade-offs between customer interests and design decisions.

We recognized that these new capabilities allowed us to really solve important company problems and improve business results. We also realized that these new tools, effectively used, could dramatically improve our new product development process. In addition to being president and CEO, I was also acting manager of product development at the time, so my role was to provide effective leadership to ensure this happened.

Once we were comfortable that our teams were relatively proficient in LAMDA and K-Brief problem-solving, I made two decisions that in hindsight were critical. First, I required that all problems be solved using LAMDA/K-Briefs — no ifs, ands, or buts. I would not discuss any problem unless it was documented in a Problem K-Brief. This forced us to adopt the training and methodology into our culture. Over the next six months, we cleared out our problem backlog without a single recurring issue. We also agreed that whenever we opened a new Problem

K-Brief, we would resource it, until it was really solved. We also decided to hold off developing new products until we cleared the problem backlog. Thus, I can testify to the effectiveness of Strategy 1 that you will learn about in Chapter 6.

My second decision was to only develop new products the new way, understand the customer interest, and only start design when all identified knowledge gaps were closed. One particular new project we were considering was an in-line testing machine for leak detection in plastic containers. I refused to let the engineers start designing until we closed all known knowledge gaps. Marketing was furious; we had a very tight schedule to meet our market commitments. Even Michael Kennedy, who was there at the time, challenged whether we were ready for this step. I told him it was about time to see whether this stuff actually worked. In reality, I knew we were going to miss the schedule anyway working the old way — based on our history. We might as well give this new way of working a shot. To drive and measure progress, management met with the development team twice per week to review their K-Briefs and discuss their progress closing knowledge gaps.

This project started with a scurry of activity to capture the true customer interests, to understand causal interactions, and to identify all the critical knowledge gaps — all of which we should have known in the past but didn't. We were always too busy designing and redesigning the product. The bottom line: we met the schedule, months earlier than the old way and with a great product that was highly successful; this success changed the company forever. So, I can also testify to the effectiveness of Strategy 3 that you will learn about in Chapter 6.

In November of 2008, three and a half years after we started, I retired from Benthos/TapTone. I have been invited back many times over the years, always finding these new processes still firmly in place and still supported by both the engineers and management. The product lines have expanded from that initial start. Not only that, the knowledge-driven processes were also successfully adopted at other Teledyne companies and since then many companies outside of Teledyne.

So why do we need another book? Why not just convince teams to copy what we did at Benthos? In many ways, that makes sense. The basics are very understandable: don't make decisions unless you understand and have resolved the ramifications and trade-offs of those decisions. However, even at Benthos, where we had fully trained and committed leadership and an engineering team that fully understood decision ramifications, it was still a challenge to learn and adapt the new practices into our culture. Now, expand leadership across many levels of management and expand a single system to a system of systems and the level of difficulty grows exponentially. How is it possible to even effectively identify all the knowledge gaps, much less understand and resolve all the trade-offs for optimizing all of the decisions?

That is what is facing product development today. That is the challenge that I saw when working with TCC (Targeted Convergence Corporation) at many organizations like Pratt & Whitney, which you will read about in Chapter 1.

The learning gained from working with dozens of similar product development organizations and the way to put these practices into action have been packed into this new book. You will read about the skills that teams will need to master in order to achieve these new practices. And more importantly, it addresses the leadership behaviors required to make it all happen.

In this book you will learn about product development tools that will be vital to your success, especially in a large organization developing complex products or systems. In smaller organizations with less complex products, this Causal Mapping will be helpful, but success is still possible using the basic tools (LAMDA, K-Briefs, and Trade-Off Charts) to achieve "Success is Assured."

Whether your company is big or small, whether your products are simple or complex, one thing is the same: the requirement to align your behavior to "True North," which you will read about in Chapter 2. OK, so what exactly is "True North" and how do you align to it? It all starts with trusting your engineers; given the proper tools and training, they will know what has to be done. All management has to do is demand to know the impact of the key engineering decisions.

Your biggest enemy is the detailed schedule created and delivered to the team at the beginning from on high. Close behind is wishful thinking; this is how most companies get through the major milestones in the product development process. It helps them avoid early scheduling delays until the problems are so catastrophic that they cost about a hundred times more to fix later in the development cycle. Why do companies make this expensive mistake over and over and over again? Do you make this mistake?

Aligning your behavior to "True North" will eliminate this problem. Basically, it is the relentless search for the truth and the willingness to act accordingly — NOW. If you are on a Toyota production line and you see a problem, what do you do? STOP THE LINE. Why? The problem will only cost much more to fix later when the warehouse is filled with scrap. Why not take the same approach in product development? Problematic knowledge gaps will only be more expensive to close later in the process.

To ensure a successful transformation, be sure to know your true customer interests, provide your engineers with the best tools available, and require that the truth always be told and that the impacts of all engineering decisions are clearly understood.

When you know exactly what to do and exactly how to do it, product development is really simple.

This is not intended to be a book of theory; it is a how-to manual. We know it works; we also know it will be hard work. Trust me, it will be worth it.

Ron Marsiglio
Teledyne Benthos

Acknowledgments

First, we would like to thank Pratt & Whitney for approving the release of the case study you'll find in Chapter 1, including a number of images that they allowed us to use. And of course, the case study would never have happened had they not had the vision to allow Penny and Brian Gracias to do what they did, including bringing in Michael and TCC for training and coaching.

Any artistic quality that you find in the figures in this book can be attributed to Kent Harmon and Caitlin Kennedy. Beyond that, though, Kent Harmon is actually a major contributor to all of the content in this book, as he has developed much of the training material that TCC has used over the last dozen years, and this book was largely derived from those training efforts. We cannot overstate the contributions that Kent has made.

Next, we need to thank all of those workshop participants over the years that pushed back on us, forcing us to improve how we teach, how we coach, and how we mentor. It is not easy when you are trying to transition an organization to a completely different paradigm — not easy for us and not easy for our partners and champions at those companies. We can't possibly name all those workshop participants, but many of them will appear in the list of people we asked to review this book.

Agreeing to review early drafts of a technical book like this is a major sacrifice, and we were very fortunate to have so many who agreed to do so:

Bella Englebach (Janssen), Betty Kennedy, Bob Melvin (Teledyne Marine), Brian Gracias (Pratt & Whitney), Bruce Newman (Siemens Healthineers), Cathal Flanagan (Graham Packaging), Charlie Camarda (NASA), Clint Carter (TCC), Colin Gilchrist (Fisher & Paykel/Set Based Solutions), Dave Zilz (Boeing), David Cole (AutoHarvest), David Thompson (Siemens Healthineers), Dean Mueller (Boeing), Dick Gall (Crane), Dick Weber (L3 Technologies), Donny Blair (L3 Technologies), Durward Sobek (Montana State University), Göran Gustafsson (Chalmers), Greg Tucker (Raytheon), Håkan Swan (Ivolver), Ian Jensen (AIAG), Jason Irby (TCC), Jim Eiler (TCC), John Leuer (Boeing), Luc Delamotte (Thales), Manoel Santos (Embraer), Marc Nance (Boeing), Mary Poppendieck, Michael Buckley (Humanproof), Mike Jones (TCC), Norbert Doerry (NAVSEA Naval Sea Systems Command), Ovidiu Contras (Bombardier), Peter Palmér (Scania), Ramakrishnan Raman (Honeywell), Rich Gildersleeve (DJO), Rob May

(Bombardier), Ron Marsiglio (Teledyne Benthos, KnowledgePD), Scott Gray (AIAG), Steve Scotti (NASA), Tim Bridges (Boeing), Tom Cloft (Pratt & Whitney), Tom Poppendieck, and Vasilije Drecun (Siemens Healthineers).

Of those, there were a few that actually reviewed multiple revisions of this book and/or gave us significant amounts of feedback that have resulted in dramatic improvements. We have to give a special thanks to them, as that was way more than we should have asked for!

Brian Gracias (Pratt & Whitney), Bruce Newman (Siemens Healthineers), Charlie Camarda (NASA), Clint Carter (TCC), Colin Gilchrist (Fisher & Paykel/Set Based Solutions), Dave Zilz (Boeing), David Thompson (Siemens Healthineers), Dean Mueller (Boeing), Dick Weber (L3 Technologies), Donny Blair (L3 Technologies), Durward Sobek (Montana State University), Göran Gustafsson (Chalmers), Håkan Swan (Ivolver), Luc Delamotte (Thales), Manoel Santos (Embraer), Marc Nance (Boeing), Norbert Doerry (NAVSEA), Ovidiu Contras (Bombardier), Ramakrishnan Raman (Honeywell), Rob May (Bombardier), Ron Marsiglio (Teledyne Benthos, KnowledgePD), Tim Bridges (Boeing), Tom Cloft (Pratt & Whitney), and Vasco Drecun (Siemens Healthineers).

Finally, we need to give special recognition to one workshop participant (and book reviewer): Ron Marsiglio. We push all the executives who participate in our workshops to "be like Ron"… to push aside all the competing initiatives and politics and keep the focus squarely on what's most important. Since that successful transformation at Teledyne Benthos, Ron has worked with us at several of our clients, sharing his perspectives and insights — including at Pratt & Whitney, as you'll read in Chapter 1. And Ron was kind enough to share his perspective on that role in the Foreword to this book.

Introduction

Why Did We Write This Book?

To answer this question, we first need to give you some context.

Everyone has heard of Wilbur and Orville Wright's successful powered manned flight in 1903. But do they appreciate why the Wright brothers succeeded when many other more educated and well-financed individuals failed?

The nineteenth century was littered with attempts to achieve manned flight, with often hilarious and sometimes fatal results. One of the more accomplished was Otto Lilienthal of Germany, who developed many successful gliders and published widely used data on lift and drag characteristics. Unfortunately, he died in 1896 during one of his flight tests.

In the United States, talented and award-winning scientist and engineer Samuel Langley began experimenting with powered flight in 1887. He tested dozens of model airplanes. Based on the results, he was awarded nearly $70,000 in grants from the War Department, the Smithsonian, and others to develop a manned airplane. Even with these resources, Langley was never able to learn enough to design a working manned airplane.

In contrast, Orville and Wilbur Wright, lacking engineering degrees or even high school diplomas, had little aeronautical experience and essentially no budget. However, they did know how to build machines quickly as they owned a bicycle shop. We all know they succeeded in achieving powered manned flight where everyone else up to that point had failed, and they did it with a fraction of the resources and in much less time. How did they do that?

The Wright brothers began like most of the others: researching the available knowledge on the topic and then designing and building some gliders to test. From that testing they found that their results did not match Lilienthal's, and they couldn't explain the differences. These early efforts led Wilbur to make the assertion that "men would not fly for fifty years."[1] In other words, they were on the same trajectory as everybody else as long as they followed the same practices as everybody else.

---

[1] M.W. McFarland (editor), *The Papers of Wilbur and Orville Wright*. New York, NY: McGraw-Hill, 1953, p. 934.

However, in 1901, the Wright brothers broke from that common pattern of "design then build then test and repeat." They stopped designing new aircraft and instead focused on closing the key knowledge gaps that prevented them from knowing the right design decisions to make, as they reported in a September 1901 speech in which they identified the three key knowledge gaps:

- "the construction of the sustaining wings"
- "the generation and application of the power required to drive the machine through the air"
- "the balancing and steering of the machine after it is actually in flight"[2]

To close the knowledge gaps regarding the design of the wings, they designed and built a wind tunnel (Figure I.1) and delicate scales (Figure I.2) that allowed them to take accurate measurements of the effects of hundreds of different wing designs in just a few months.

They learned how wingspan, wing chord, angle of incidence, aspect ratio, surface area, and airfoil shape each independently affected the lift and drag, and how various combinations worked together. They created charts of that data (Figures I.3 and I.4), which gave them visual insight into the cause-and-effect relationships across the whole design space.

This learning turned out to be immediately reusable as it helped the brothers design an aerodynamically efficient propeller, in contrast to the marine-inspired propellers used by their predecessors. Knowing the limits of their wings and their propeller, they were able to compute that they only needed an 8 hp engine, allowing them to use a very simple 12 hp engine design cast in aluminum to make it lighter in weight. (In contrast, without that knowledge, Langley spent the bulk of his budget developing a very sophisticated lightweight 70 hp engine.)

Figure I.1   Wright Brothers' wind tunnel (replica).

[2] M.W. McFarland (editor), *The Papers of Wilbur and Orville Wright.* New York, NY: McGraw-Hill, 1953, p. 99.

Figure I.2   Wright Brothers' balance scales.

To verify that their wind tunnel measurements on tiny wings would translate as expected to full size and to test their control mechanism designs, they built a full-size glider that they could operate as a kite for the safety of the pilot. Once they had adequately optimized the design to make control easier for the pilot, they began making untethered glides of several hundred feet. During September and October of 1902, they made 700–1000 test flights, verifying there were not any unexpected control issues in different conditions.

With all three knowledge gaps closed, they then designed and built their powered manned flying machine. Their first test failed as the pilot (Wilbur) overcompensated in the use of their new control system. But in that failure, they demonstrated that they had closed the last of the knowledge gaps (even with the weight of the engine and the gyro effects of the propellers, they still had more than enough control). After this failure, Orville confidently sent a telegram to his father: "Misjudgment at start reduced flight to hundred and twelve power and control ample rudder only injured success assured keep quiet."[3] Given everyone else had failed repeatedly, claiming "Success is Assured" based on a test flight failure is extremely bold — unless you have knowledge that proves it is so.

---

[3] M.W. McFarland (editor), *The Papers of Wilbur and Orville Wright.* New York, NY: McGraw-Hill, 1953, p. 393.

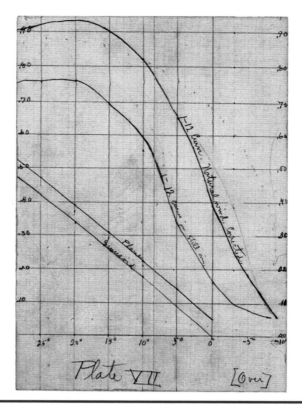

Figure I.3   Trade-Off Chart example from the Wright Brothers' notebooks.

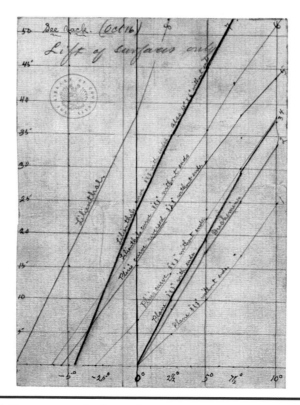

Figure I.4   Trade-Off Chart example from the Wright Brothers' notebooks.

Figure I.5 **The first successful flight of a manned, powered, heavier-than-air vehicle.**

Three days later, with no design revisions (just an elevator repair), the Wright Flyer (Figure I.5) flew 852 ft in 59 seconds. With their approach to product development, their first powered manned airplane design succeeded. It took them 22 months (spread over four years), with a budget of under $1000 and a staff of three. In contrast, Langley generated numerous costly failed designs over the course of 16 years, spending well over $70,000 and putting his test pilots at great risk.

The key takeaway: The contrast was not so much the individuals, but the process. The Wright brothers' success was due to their switch from the traditional design–build–test process to a distinctly superior process. **They identified the knowledge they needed to succeed and found innovative ways to acquire that knowledge. They didn't try to design their aircraft until their knowledge gaps were closed**. In contrast, Lilienthal and Langley used expensive prototypes to identify their knowledge gaps and then modified those prototypes based on their new knowledge and tried again. This resulted in slow, expensive learning cycles and in the end did not result in a working design.

So, With That Introduction, Why Did We Write This Book?

Because of two very different shared experiences...

The first is an experience we share with almost every veteran of complex product development that we meet: the same frustrations with the same processes due to the same underlying problems. Despite decades of improvement initiatives, the results remain stubbornly the same.

- Delays and missed milestones, poor time-to-market
- Rampant firefighting and costly rework
- Budget overruns
- Missed revenue and/or profitability goals
- Quality and customer satisfaction issues

And in many ways, things seem to just get continually worse. The blame for this is usually directed at ever-increasing complexity and ever more unreasonable demands. Organizations are being squeezed on all sides.

- More cost-cutting measures
- More regulations
- More requirements
- More electronics
- More software
- More specialization
- More expertise boundaries
- More global competition

The second is an experience that we share with a much smaller group of people: people who understand the learning from the Wright brothers and have seen the impact of actively trying to establish "Success is Assured" for the critical decisions that they make. That impact is far reaching.[4]

- Two- to fourfold increases in productivity
- Development cycle times cut in half
- Much higher schedule reliability
- Improved quality and innovation
- Rework almost entirely eliminated
- Effective knowledge reuse and continuous improvement
- Consensus decision-making based on knowledge
- Shorter and more effective meetings
- And so on

With this book we want to share our second experience with all of those with whom we currently only share the first experience.

The Root Cause

Failure to Address the Root Cause

In general, the product development improvement initiatives companies have used for years are well intentioned, but they do not address the underlying root cause responsible for these undesirable results. Instead, they focus on schedule adherence and task execution. More oversight is added to manage schedules, and task execution instructions are created to improve execution results. In Chapter 1, we will elaborate on the failure of companies' improvement initiatives to address

---

[4] M.N. Kennedy, J.K. Harmon, and E.R. Minnock, *Ready, Set, Dominate: Implement Toyota's Set-Based Learning for Developing Products and Nobody Can Catch You.* Richmond, VA: Oaklea Press, 2008.

the real root cause with a real-world example: despite a long history of improvement initiatives, Pratt & Whitney realized they needed a different way of thinking to enable better product development performance. Their journey, while not yet complete, will hopefully inspire leaders to look more deeply at the opportunities to improve their product development systems.

Missed milestones and poor task execution are symptoms of the root cause, not the real root cause.

So, What Is the Root Cause?

The real root cause is making critical design decisions early in the development process before the required knowledge is available, resulting in decisions that routinely need to be remade, in turn resulting in tremendous rework and churn.

Figure I.6 may look familiar. During the early concept and preliminary design phases, many aspects of the design remain fuzzy (enough so that the term *fuzzy front end* is commonly used to describe it). As the design continues through the detailed design phase, the design details can then be accurately analyzed and tested. This exposes that the design will not adequately satisfy the requirements. That realization then drives design changes, often to decisions that were made during the fuzzy front end. The result is much of the detailed design work based on those early decisions may need to be redone — very expensive rework.

Knowing the amount of learning that occurs late in the development processes, management often responds by trying to accelerate the front end of the process to get to the back end learning sooner. Some improvement initiatives actively promote such schedule compression in the hope of getting to the learning (via firefighting) sooner. However, in doing so, their schedules end up applying additional pressure to make decisions early, which results in more

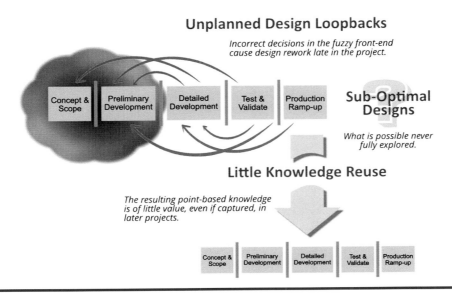

Figure I.6   The typical wastes in the fuzzy front end of development.

decisions being made prior to having the required knowledge, exacerbating the root cause.

In addition to the resulting bad decisions, the new knowledge acquired as part of the late process loopbacks is typically left poorly documented because of the chaotic firefighting needed when the project is late and over budget. Even when documented, it tends not to be generalized beyond the specifics of that project (that specific point design) and thus not applicable to the next. The result: the same poor decisions are made again in the next product development program (at a different point in the design space).

In response, companies also use another mechanism to deal with this: "Risk Management." They identify the risks they are accumulating as the process moves forward (a good thing). In some cases, they actively work on eliminating those risks (a very good thing), but too often they decide that the risks are unavoidable and need to be accepted as risks so that the project can move forward. When they accept those risks as unavoidable, they will typically mitigate them by adding schedule buffers or by adding design margins; the first leads to even more delays, the second leads to suboptimal designs. Worse, such risks are not just additive but tend to multiply and compound, particularly in complex projects. Almost inevitably, all the tidy schedules and plans from the front end planning become chaotic as the engineering teams firefight their way to success through late problem-solving.

In their book *The Minding Organization*,[5] Dr. Moshe Rubinstein and Dr. Iris Firstenberg predicted as much when they stated that in a creative environment, if you begin with deliberate order, you will move to emergent chaos. In other words, you are fooling yourself if you believe you can plan your way up-front through a minefield of knowledge gaps. Ironically, the learning itself in a focused firefighting mode is quite efficient, as the teams are focused and committed — even though the rework is pure waste.

Rubinstein and Firstenberg also predicted that if you begin with deliberate chaos, you will tend to move to emergent order. Think about it: if you begin with efficient, rapid, collaborative learning to seek out the knowledge gaps, then your teams can build and execute urgent but workable plans to eliminate those gaps without a plethora of ongoing surprises.

That is what this book is about: **We want to put in place up front a structured and efficient convergent learning process that is high energy, adaptive, and collaborative. A key goal is to eliminate as much risk as possible and minimize the need for ongoing risk management.** This will reduce the chaotic, expensive, and wasteful firefighting at the ends of these processes. Various studies have shown that the cost of remaking a decision late in the process is 100 to 1000 times more expensive than it is to make the original

---

[5] M.F. Rubinstein and I.R. Firstenberg, *The Minding Organization: Bring the Future to the Present and Turn Creative Ideas into Business Solutions.* Hoboken, NJ: Wiley, 1999.

decision.[6] So, elimination of even a small amount of rework can justify a tremendous amount of learning up front.

Rework is a very slow, laborious, and expensive way to do that learning. When focused on learning first, the team is performing experiments crafted to inexpensively create knowledge. The alternative is behaving as if we knew enough to make the final design decisions and thus doing all the detailed work that must be done to get a product out the door.

Even when faced with things that cannot be fully learned up front such that risk must be taken on and managed, by fully understanding the potential impacts and learning the full design space around those impacts, you are maximizing your flexibility to quickly respond to the learning that comes later in the process. Keeping multiple options open as late as possible within the overall design space can mitigate the impact of late-process learning.

The organization that learns the fastest will ultimately "win." They will eventually pass their competitors that are learning more slowly, and those competitors will then be constantly trying to catch up. Development processes should be judged primarily on how rapidly and efficiently they build reusable knowledge that enables the decision-making that they need to get their products out the door.

The Solution: Establish "Success is Assured" before Making Critical Decisions

To address the root cause above, the decision-makers need to resist making decisions until they have knowledge that establishes that "Success is Assured." We will elaborate more on that mandate in Chapter 2.

At first glance, you might say "Well, that's kind of what our phase gate process requires." Such language *is* often in one or more of the gate criteria, but let's look at some common behaviors. For example, we usually have a risk management process in place. We roughly identify the different ways we might fail and then try to put in place appropriate measures to prevent or to react to such failures if they should occur, along with some padding in the schedule and budgets to deal with the fallout. **That is not "Success is Assured"; that is a guess at how bad our guess is**.

Striving for "Success is Assured" calls for you to identify the knowledge gaps that are the root causes of that risk and then actively close the knowledge gaps to *eliminate* the risk. While it might seem more efficient at the time just to push the risk downstream and deal with it later, the reality is the risks will generally compound as time goes on.

---

[6] INCOSE, *INCOSE Systems Engineering Handbook: A Guide for System Life Cycle Processes and Activities*, San Diego, CA: International Council on Systems Engineering, 2011; credited to Defense Acquisition University, 1993.

You will know you have moved to "Success is Assured" practices when

- Risk management is largely replaced by *knowledge gap closure* (risk elimination) and *Set-Based knowledge* (risk mitigation via flexibility to handle the worst cases in the set).
- Debates in meetings on critical decisions are largely replaced by consensus decision-making based on shared visual knowledge that makes the right decision clear to everybody.
- Very little focus is given to gate criteria verifying that specific tasks have been completed; rather, the focus is on the knowledge justifying the design decisions that have been made and the trade-offs being made on behalf of the target customer(s).
- You don't find yourself repeating the same mistakes made in previous programs; your decision-making process consistently leverages the knowledge from past programs.
- You see higher levels of innovation, all focused precisely where you need it, finding ways to better assure success within the required timeframes.

Easier Said than Done: The Enablers

Of course, closing such knowledge gaps may take time, so striving for "Success is Assured" often requires delaying the decisions to make time to close the knowledge gaps. But some of the testing and analysis to close the knowledge gaps requires you to know the decisions that haven't been made yet. A catch-22?[7]

That's where **adoption of Set-Based practices becomes a critical enabler**. The Set-Based paradigm replaces traditional Point-Based practices with a convergent design discovery process. Early in product development, many things are fuzzy, many decisions are not yet made, so analyses and testing will need to be done for sets of possibilities to gain visibility of the different trade-offs available in the full design space. **Decision-making will not be trying to "pick the best" but rather to "eliminate the weak," steadily converging to the final answer as more is learned about the narrower design space, as more decisions are converged**.

Furthermore, in complex system design, those catch-22 circular dependencies typically cross boundaries of expertise. So, the Set-Based analyses need to be conducted collaboratively. Even identifying the knowledge gaps that need to be closed may not be possible without experts with several different specialties collaborating effectively. It becomes essential to have a collaborative method that enables a variety of experts to identify the *unknown unknowns* (the knowledge

---

[7] "Catch-22" is English slang for a paradoxical situation in which the mutual dependencies make escape or resolution seemingly impossible.

gaps that you don't yet know you have), uncover the necessary knowledge to begin "eliminating the weak," and then help them steadily converge a set of decisions to a portion of the design space for which "Success is Assured."

We will elaborate more on those enablers of the "Success is Assured" mandate in Chapter 3, and introduce a specific collaborative methodology for the latter in Chapter 4 that we call "Causal Mapping." This can be applied to traditional problem-solving scenarios (where you don't want to cause new problems when fixing the old) or to large complex system-of-systems design challenges with multiple objective criteria requiring expertise from multiple disciplines.

Applying This Across Organizations

In complex product development, it is not enough to focus your own organization on establishing "Success is Assured" and to put in place the enablers to get there; you must work with other organizations that may or may not have the same focus and the same capabilities. Chapter 5 discusses methods to align these efforts across multiple organizations in a way that allows them to operate independently and concurrently but still in a coordinated way.

Making the Transition

We conclude Part I of this book with some recommended strategies for making the transition from your existing practices to the new practices, depending on the situation in your organization. We discuss some of the common failure modes and obstacles that you should be watching for and reacting to. Note that senior leadership may play a critical role in overcoming some of the key obstacles facing their organizations trying to transition to these practices. That is all discussed in Chapter 6.

Putting It All Together

There are quite a few enablers for establishing "Success is Assured" (to be detailed in Chapter 3) and each depends on the others. It is often hard to appreciate how they enable each other until you apply them to your real-world situations. But it's hard to get people to adopt new practices until they see them work. So, how do we share with you our shared experiences in establishing "Success is Assured"?

Our answers are Chapter 1 and Part II (Chapters 7–16) of this book. Chapter 1 tells the story of two engineers' grassroots journey to introduce "Success is Assured" into their product development process. The hope is that you will see your company's product development evolution in the Pratt & Whitney story and have some ideas about how to establish "Success is Assured" within your process. The goal of Part II is to let you sit in on a series of design meetings very early in a product development effort and experience what we see over and over when we coach teams through such meetings. We simply apply those enablers as the team discusses the problem, and those enablers help them identify the knowledge gaps that need to be closed to make the decisions that need to be made.

Although Part II is a fictional set of meetings, every piece of those fictional meetings is something we have heard or experienced in real-life meetings. You essentially get the best-of-the-best "aha"s rolled into one cohesive story, hopefully accelerating your sharing of all those "Success is Assured" experiences.

In our work with numerous companies, only a small percentage of product developers (whether engineers or managers or otherwise) are able to "get it" from just our basic training on the method. It is not until they experience its application to their own product development challenges that they truly "get it." And since complex product development is inherently a collaborative activity, doing so presents a challenge. You need the rest of the team to be applying the new methodologies *together* for them to experience the effects and "get it."

So, it is our hope that Chapter 1 and Part II together will enable your team members to each experience it enough that they adequately "get it" such that they can successfully apply the methodologies in their own work. But we also remain available for coaching support to whatever extent your organization needs. Drop us an email.

Our Target Audience

This book is primarily targeted at product development organizations, including the engineers and other subject matter experts, project management, middle management, senior leaders, and executives. They all have important roles to play in making the transition to superior practices. And they all must understand how their role fits in with the other roles. In the end, major improvements to product development performance is only possible by changing how those organizations make the complex decisions that they need to make. Anybody involved in those decisions, whether making those decisions or providing the knowledge to make those decisions, needs to understand the new decision-making process. And given we will be providing visual models that tie the knowledge to the decisions such that all can evaluate whether the knowledge proves that "Success is Assured," all involved need to understand those visual models.

I Am Not in Product Development, but I Do Make Complex Decisions

The methodologies and tools presented in this book are actually designed **for any complex decision-making situation**. And some of the organizations we work with have applied them to scheduling problems and business problems. However, rather than trying to generalize beyond the bulk of our experience, we have chosen to stick to examples and language oriented for product development. But you can feel confident that you are not going into unchartered waters if you plan to apply this to other complex decision-making or complex problem-solving scenarios.

HOW TO ESTABLISH "SUCCESS IS ASSURED" PRIOR TO MAKING DECISIONS IN YOUR ORGANIZATION

I

Part I begins with a case study of Pratt & Whitney and then explains the lessons learned from our work with dozens of similar organizations, refined into clear how-to instructions for an organization's decision-makers at all levels.

Chapter 1

Pratt & Whitney Case Study: Their Journey to "Success is Assured"

The rest of this book will give specific how-to advice that you can implement in your organizations. To set that up and sensitize you to the challenges you will likely face in initiating such changes, this chapter will give concrete examples of trying to make such improvements within a complex organization. It will allow you to see both what worked well and what did not work so well. It will also illustrate the importance of many of the elements that will be described in the following chapters (as many of them were inspired by experiences such as these).

Further, in most of the organizations where the content of this book has been taught, there are at least a few people that have trouble focusing on the recommendations because they see so many obstacles to their adoption in their organizations. Often, the less vocal people will also share that they too were highly skeptical this would actually have an impact due to those obstacles. One of the most common questions is

Where has this been done successfully in a company like ours?

It seems the vast majority of people don't want to be first adopters. And they don't want to waste their time learning new skills if their organization's management or situations will end up preventing those skills from being used.

So, to allow readers to focus on the teaching in the rest of this book, this chapter will try to address this question up front. Obviously, it can't answer this for your specific company, but it can walk through a case study of a company that has many, if not most, of the challenges and obstacles that you are concerned with in your own organization. And even with such obstacles leading to only partial adoption (so far), the effort has had significant value and they continue to be interested in progressing further.

This case study will attempt to make clear that the techniques taught in this book are applicable in the real world, that they can be applied despite the obstacles, and that the skills you learn will make you and your organization more effective, even if there is only partial adoption.

The Setting: Pratt & Whitney

Pratt & Whitney (P&W) is a world leader in the design, manufacture, and service of aircraft engines and auxiliary power units, and has been powering aircraft for over 90 years. Founded in Connecticut in 1925, P&W's first aircraft engine — the air-cooled Wasp (Figure 1.1) — transformed the aviation industry with its unprecedented performance and reliability.

P&W has been leading change ever since. Today, P&W develops game-changing technologies such as the PurePower® PW1000G engine with patented Geared Turbofan™ engine technology. Its large commercial engines power more than 25% of the world's mainline passenger fleet, and its military engines power frontline fighters and transport aircraft for 31[1] armed forces around the world.

So, P&W is a very large company, with more than 38,000 employees worldwide and $15 billion in adjusted net sales (2016). In that sense, P&W serves as a good example given that adopting change can be much more challenging in large, well-established companies.

Even more challenging, P&W is just one of four business units of the larger United Technologies Corporation (UTC; NYSE: UTX), which serves customers in the commercial aerospace, defense, and building industries and ranks among

Figure 1.1   First P&W Wasp engine built.

[1] Pratt & Whitney, Military engines. Online: http://www.pw.utc.com/Military_Engines (accessed October 28, 2016).

the world's most respected and innovative companies. UTC employs more than 200,000 people in more than 70 countries worldwide and has adjusted net sales of $57 billion (2016). It is ranked 45 in the Fortune 500. The other three business units are Otis, UTC Climate Controls & Security (key brands: Carrier, Chubb, Kidde), and UTC Aerospace Systems (key brands: Hamilton Sundstrand & Goodrich Aerospace).

P&W develops and manufactures a fairly large number of very complex products (jet engines and auxiliary power units) that have extremely high reliability requirements and yet often need to push the boundaries on performance. Table 1.1 shows a sample of those products and the aircraft they power.

Table 1.1   Sampling of P&W Engine Models

	Engine Model	Fan Diameter (in.)	Maximum Thrust (lbf)	Fan Bypass Ratio	Planes Powered	Engines per Plane	Entry into Service	Major Partners
Commercial engines	JT8D	39.9–49.2	39,900–49,200	1/1–1.7/1	Boeing 707, 727, 737 DC9[a] MD80[b]	4/3/2 2 2	Q1 1964	
	PW6000	56.5	22,000–24,000	4.9/1	Airbus A318	2	Q4 2005	
	PW4000	94–112	52,000–90,000	4.8/1–6.4/1	Boeing 747, 767, 777 Airbus A300 MD 11	4/2/2 2 3	1987	
	GP7000	116	70,000	8.8/1	Airbus A380	4	2008	GE[c]
	V2500	63.3	23,000–32,000	4.5/1–5.4/1	Airbus A318, A319, A321	2	Q2 1989	RR[d]
	PurePower® PW1000G	56–81	15,000–33,000	9.1/1–12.1/1	Bombardier C Series Airbus A320neo	2 2	2016 2016	
					Mitsubishi Regional Jet	2	2020	
					Irkut M-21 Embraer E-Jets	2 2	2018 2018	
Military engines	F100	46.5	23,770–29,160	1/1	MD[b] F15 GD[e] F16	2 1	Q1 1976 Q3 1978	
	F119		35,000		Lockheed Martin F22	2	Q3 2002	
	F135 CTOL	52	43,000	1/1	F35 Lightning II	2		RR[d]

Source: pw.utc.com

[a] Douglas
[b] McDonnell Douglas
[c] General Electric
[d] Rolls Royce
[e] General Dynamics

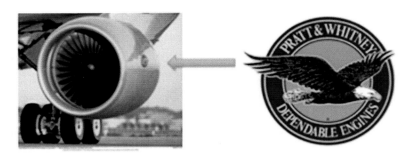

Figure 1.2 Engine with P&W eagle.

If you are on an airplane and can see the engines outside your window, look for the P&W eagle (Figure 1.2) and you will know you are flying on the most dependable engines in the world!

Establishing "Success is Assured": The Promise and Challenge

In order to establish that "Success is Assured" prior to making key decisions, an organization must be willing to spend more time learning *fuzzy knowledge* in the earliest phases of product development, delaying decision-making until that learning can be completed. Doing so can seem counter to traditional approaches that focus on leveraging sophisticated back-end simulation and/or testing capabilities to do the learning, and thus try to quickly nail down the requirements and then the specifications in the front end to more quickly get to those back-end analyses. The sooner we can get to those analyses, the sooner we can learn what is wrong (the "bugs") and focus on fixing them. However, those "bugs" can consistently be traced to key decisions that were made early in the process.

The promise of establishing "Success is Assured" prior to making those decisions is avoiding the time and money wasted in all the rework (redesigning, reanalyzing, and retesting), which can consume 60%–80% of your engineering capacity. But the costs don't end there, as production hardware is already on order and qualification testing is on a tight schedule to meet the customer deadline. To recover, hardware on order must be canceled and qualification testing delayed and rescheduled, frequently leading to expensive cancellation fees and unhappy customers. Replacement hardware is needed quickly, incurring costly expediting fees. Costs and delays just keep adding up.

Given that, establishing "Success is Assured" at P&W may seem like common sense, but its complexity is immense. Between the size of the company, government regulations, and customer requirements that often challenge the boundaries of physics, the task of facilitating change is daunting. And then there are additional complexities:

■ Materials are highly specialized to withstand the extreme environments required.

- Capable suppliers are limited and manufacturing lead times are long.
- Orders must be placed as soon as possible to meet aggressive development and production schedules. (This means that orders are placed before the preliminary design is complete.)
- Engine systems' interrelationships and their effects are hard to predict.
- Data for current and past engine configurations may not help when new technologies are introduced to meet customer requirements. Unknowns are elusive until the engine is tested.

Convincing P&W management they could establish that "Success is Assured" despite the cost and schedule pressures wasn't going to be easy. Can a company with products as complex as jet engines really learn early enough, efficiently enough, and complete enough to avoid huge redesign and analysis efforts to flesh out all the issues?

My name is Penny Cloft, Senior Fellow Discipline Lead (SFDL) emeritus. I had a partner in this journey: Brian Gracias, Engineering Quality Manager and Six Sigma Master Black Belt. Brian and I share a passion for process improvement and together led the charge to change Pratt & Whitney's product development process with the help and support of many fellow P&W employees along the way.

This case study will give you a response to those who push back because they believe your products are too complex, your obstacles are too diverse, or your organization is too entrenched in existing practices.

Read this story with an eye to adapting the principles and implementation to suit your company's situation. Keep your focus on what was done, why it was done, and how it will relate to your company. Keep in mind there are two interwoven tracks on this journey: the P&W (company) story of change and the personal learning of change agents and leaders. Change management is an important part of any improvement journey. It can be messy, particularly if the vision and goals are not absolutely clear.

Along the way I will provide you with insight into my perspective on this journey. But first I will review P&W's past continuous improvement initiatives on their product development process, as they need to be considered when plotting a path to new changes.

P&W's History of Successfully Implementing Continuous Improvement Initiatives

P&W has a very successful history of implementing continuous improvement initiatives, so much so that Jim Womack, in his book *Lean Thinking*,[2] made

---

[2] J.P. Womack and D.T. Jones, *Lean Thinking*, New York, NY: Free Press, 2003.

note of that success. However, with such success can come a lot of resistance to change. Although establishing "Success is Assured" prior to decision-making represents a dramatic change to existing successful practices, it can also be shown to be a complementary fit to the goals of most of those initiatives. Finding and making visible that alignment can be critical to overcoming the objections to change. Chapter 2 will give general recommendations on that, but this chapter will give some concrete examples for those general recommendations to be built on.

In the early 1990s, I had a manager who liked to start his meetings with a view graph stating:

Equally informed people seldom disagree.

I don't know where the phrase came from, but I enjoyed watching the reactions of the people in the room. It set the tone for respecting everyone's point of view, and people seemed more willing to share their ideas. The phrase stayed with me for the rest of my career. It changed how I approached my work and how I dealt with people. I also realized that the success of many of the changes to our product development process came from the desire to achieve that phrase.

When companies are small they can fit their employees in one building, and many of their employees perform more than one specialized function. Communication and the sharing of ideas and data come more easily than when a company is large and people are located all over the country or the world. Add to that a complex product that requires focused expertise in specific areas like noise abatement, aerodynamics, structural loads, or thermodynamics, and sharing can become impossible without a lot of intervention. A lack of that sharing can have significant negative implications on product performance, sales growth and brand reputation. At P&W, I often found myself saying, "Equally informed people seldom disagree, *but getting them informed and truly understanding is a very difficult task*!"

There are three major P&W continuous improvement initiatives in product development that are worth reviewing.

1. Integrated Program Deployment (IPD)
2. Lean Manufacturing
3. Phase Gate product development process and the implementation of a Systems Engineering organization

These initiatives all had something in common: they strove to improve quality and lead times and reduce costs by improving communication (getting people equally informed) and focusing on the customer. Unfortunately, they are not sufficient to establish "Success is Assured."

Initiative 1: Integrated Program Deployment

In the early 1980s, P&W began to face strong competition from General Electric (GE) and Rolls Royce (RR) as they expanded their products into commercial markets. The result reduced P&W market share and profits.

To cut costs and stay competitive, P&W had to make changes. One of their solutions was to implement IPD. IPD is "a management technique that simultaneously integrates all essential acquisition activities through the use of multidisciplinary teams to optimize the design, manufacturing and supportability processes." IPD facilitates meeting cost and performance objectives — from product concept through production, including field support. One of the key IPD tenets is multidisciplinary teamwork through Integrated Product Teams (IPTs).[3] Designs reflected all discipline capabilities, thus eliminating the time wasted and quality issues experienced without their early input.

The PW4084 engine was the first engine development program to use IPD. The program saw a 1-year reduction in development lead time from five to four years. Engineering charged hours also decreased by a similar amount.[4]

IPD moved the company in the right direction. The process mandates that disciplines interact before design decisions are finalized, but there was no requirement to determine "Success is Assured" before making the decisions. The push to quickly select the design configurations was still the key driver.

Initiative 2: Lean Manufacturing

Jim Womack's book *Lean Thinking* describes P&W's Lean manufacturing journey at length. The book reviews the Lean journey of three American companies. Jim used P&W as his acid test example: "If Pratt can apply these principles quickly in a massive, publicly traded, high-tech organization with extraordinarily deep technical functions and life or death demands on the product quality… then literally any American firm can."[5] Continuous flow principles were applied to all the factories. To fully implement continuous flow, product engineers had to be part of the effort. Parts and assemblies must be designed or modified to meet the cellular manufacturing and assembly needs. To achieve this quickly and sustain the effort, engineers were relocated to the factories and aligned organizationally with the factory management and their goals. As a result, most of engineering became decentralized. Groups that focused on the entire engine, such as performance analysis, aerodynamics, system structural analysis, and specialized sys-

---

[3] Department of Defense, DoD integrated product and process development handbook, August 1998, p. 1. Online: http://docplayer.net/6044397-Dod-integrated-product-and-process-development-handbook, accessed 4/29/18.

[4] J.P. Womack and D.T. Jones, The acid test, in *Lean Thinking*, New York, NY: Free Press, 2003, Chapter 8, section "Lean but Not Lean; Necessary but Not Sufficient," p. 165.

[5] J.P. Womack and D.T. Jones, The acid test, in *Lean Thinking*, New York, NY: Free Press, 2003, Chapter 8, p. 153.

tem design engineers, remained centralized within and reporting to Engineering Management.

So what does this have to do with product development? Engineering senior management was very concerned with P&W's ability to develop new engines in a decentralized environment. When engineering was centralized, conflict resolution and technical oversight were handled completely within engineering. Now operations was in the mix. Conflict resolution and component integration could become convoluted and susceptible to influence by operational requirements while diluting engineering or the customers' requirements. To address this concern, Engineering made organizational changes. They strengthened the Engineering Management teams' ability to integrate all the modules and provide technical oversight across the factories by creating dedicated expert Component Engineer and Program Chief Engineer (PCE) positions. These experts were members of the Engineering Management team with responsibility for managing module integration and conflict resolution.

From 1993 to 1995, P&W saw production lead times reduce from 18 to 6 months[6] and reversed losses of $262 million to being in the black by $530 million. Lean manufacturing helped the bottom line, but its effects on product development were still untested.

Initiative 3: Phase Gate Product Development Process and the Implementation of a Systems Engineering Organization

By the late 1990s, Lean manufacturing and a better economy provided P&W with the cash flow to develop a new engine. P&W wanted to get back into the commercial small engine business because the market was moving away from the large airplanes for domestic flights to smaller planes. Their only small engine was the V2500,[7] which was a partnership. The V2500 was very successful, but P&W's share of the profits was small. They wanted their own engine, but they didn't want the engine to compete with the V2500 engine, so they targeted the A318 series Airbus airplanes. GE's exclusivity contract on 737s limited P&W's options. The A318 is designed to carry fewer passengers and fly shorter distances than the A320, and it competes directly with the Boeing 737-600 and GE's CFM56-5B engine.[8]

Smaller aircraft make multiple daily flights, incurring higher maintenance costs and durability issues. To become the engine of choice, P&W's engine needed

---

[6] J.P. Womack and D.T. Jones, The acid test, in *Lean Thinking*, New York, NY: Free Press, 2003, Chapter 8, section "The Bottom Line in Physical Production," p. 183.

[7] International Aero Engines AG (IAE) owns the V2500 engine. In the 1990s, IAE's shareholders were comprised of P&W, Rolls Royce, Japanese Aero Engines Corporation, and MTU Aero Engines GmbH. IAE was created so all the shareholders shared the risks and revenue of the engine. Because P&W was one of several partners, they only received a share of the revenue. If they created an engine to compete with the V2500 on the Airbus A320, they would cut into their current V2500 revenues.

[8] The Airbus A318 has seen modest market success since its introduction in 2003. Military Factory, Airbus A318. Online: http://www.militaryfactory.com/aircraft/detail.asp?aircraft_id=1034 (accessed June 7, 2016).

to be significantly better in maintenance costs and durability issues than GE's engine. This new engine, the PW6000, would be the first completely new engine to test IPD and the engineering organizational changes resulting from the Lean Manufacturing initiative. The engine was designed with fewer parts to simplify maintenance. More emphasis was put on less exotic, inexpensive materials and more robust part designs to improve durability. During engine certification, P&W discovered that the trade-off between fewer parts in the compressor and the required work output to meet performance requirements didn't go as planned. The compressor missed the efficiency targets, causing the engine to significantly miss fuel burn requirements.

Motoren- und Turbinen-Union GmbH (MTU) was developing a promising new compressor, so P&W turned to them to redesign the compressor for the PW6000 engine.[9] The new compressor was successful and the engine was certified. Unfortunately, P&W's compressor design issues caused significant cost overruns and delayed entry into service. In the end, the final PW6000 engine wasn't significantly better than its competition, resulting in very few engine sales.

So what went wrong? How could the performance miss go undetected until certification testing? The PW6000 development program was executed using IPD and the strengthened Engineering Management team to integrate the modules. These initiatives were supposed to improve communication and module integration.

After the engine was finally certified, the company performed a thorough investigation to determine the root cause and implement corrective actions. At the time, P&W was not the only company looking to improve their product development process. Other aerospace companies, defense contractors, medical device companies, and the defense department (to name a few) were experiencing similar issues. These companies were working together to share experiences and best practices, which led to the formation of international councils to collect and communicate methods that continually strive to improve product development and related processes.

Here are a few of the councils and their recommended changes to address the product development process issues that came out of this effort.

- The International Council on System Engineering (INCOSE) was founded in 1990 to improve the systems engineering discipline. Companies were finding that they lacked the discipline to think in terms of a complete, complex system. INCOSE members work together to advance the system discipline, share experiences, and create scalable solutions for system issues. Companies like Boeing and General Dynamics were advocating the use of systems engineering to improve product performance and meet customer requirements.

---

[9] MTU Aero Engines, PW6000. Online: http://www.mtu.de/engines/commercial-aircraft-engines/narrow-body-and-regional-jets/pw6000 (accessed February 9, 2017).

■ The Project Management Institute (PMI) was founded in 1969 to advance the project management discipline. The aerospace, construction, and defense industries identified the need to improve project management processes for complex products. By 2000, they were writing abstracts that encouraged the use of phase gate processes as a project management tool.

■ While not an international council, the U.S. military started using a form of the phase gate process in the early 1960s. After demonstrating success reducing program costs, the phase gate process solidified in the early 1970s. "NASA and the DoD contributed to making this approach a de facto standard by incorporating this model into the bidding process."[10] Suppliers not using a phase gate process to manage their military programs couldn't do business with the military. P&W thus used the military phase gate process on all military programs.

P&W subsequently adopted many of these recommendations to address the root cause of the issues they confronted in the PW6000 development program.

In 2003, P&W required all engine programs, whether military or not, to use the phase gate process. Phase gate processes require senior management reviews at specific time points in the development process. The intent is to ensure that programs do not proceed if they are not meeting requirements. Figure 1.3 shows P&W's phase gate process.

0	1	2	3	4	5
Concept Initiation	Concept Optimization	Preliminary Design	Detailed Design & Initial Verification	Verification & Validation	Delivery, Service & Support

Figure 1.3   P&W phase gate process. (Image: Pratt & Whitney.)

Several subsystem reviews are required within the overall phase gate process. First part and module reviews occur near the end of the phase. Missed requirements and risks are passed to the Systems Engineering organization. Then a systems-level review occurs with the engineering executive team.

A new Engineering Vice President (VP) was hired to implement a Systems Engineering organization and the Phase Gate Process. Within the Systems Engineering organization, System Design Engineer and PCE positions were created to manage system requirements and module integration. System Design Engineers create the overall systems structure, integrating the component design, airframe, and performance requirements. The PCE is responsible for all technical aspects of a product, from cradle to grave, including

■ Understanding the "Voice of the Customer"
■ Managing component integration to ensure customer requirements are met

---

[10] S. Lenfle and C. Loch, "Lost Roots: How Project Management Came to Emphasize Control over Flexibility and Novelty," *California Management Review*, vol. 53, July 2010, p. 14.

- Leading the System Phase Gate reviews and all associated activities
- Supervising the system design engineers for the program

The new VP of Engineering also implemented a Technical Fellows program to enhance the importance of discipline experts. He also added Engineering Standard Work (ESW). Experts in each technical discipline wrote standard work detailing the methods and design limits to be used when designing specific parts and components, which included

- Required inputs and outputs to perform the design or analysis
- Proven design limits not to be exceeded without management permission and a risk mitigation plan
- Proven analytical tools and methods

The intent of these additions was to strengthen the discipline focus without disrupting the existing customer and component focus of the Module Center organizations.

At all phase gate reviews (from parts, module assemblies, and system reviews), engineers must demonstrate adherence to ESW, including design standards and limits and the use of proven analytical tools. Failure to complete all tasks spelled out in the ESW could cause a failed gate review.

P&W also created a lessons-learned database to capture and store information on past engineering issues and events. The expectation was that the use of the database would reduce repeat issues experienced by prior designs.

So P&W's solution to the PW6000 was to keep the IPD process and add more oversight with the phase gates. To improve module integration they added a Systems Engineering organization. However, there was still no requirement to determine "Success is Assured" before making the decisions. The push to quickly select design configurations remained the key driver.

One of the most common objections the authors hear at the start of transformation efforts is, "How is that different from what we are already doing? Why isn't what we are doing good enough?" Hopefully, this review of P&W's product development will help you to take a similar critical look at your own initiatives. You should look for two things: (1) how the initiatives are failing to establish "Success is Assured" prior to decisions and (2) how establishing "Success is Assured" prior to decisions will actually better support your goals for those initiatives. In many cases, it could be a focus on decisions or innovative ways to accelerate learning that may help you take your existing initiatives to new heights.

Food for thought: many companies have a metric called *Cost of Poor Quality* (COPQ). What if we could add the metric *Cost of Bad Decision-Making*

(COBDM)? This would be a great way to put more focus on decisions instead of tasks and provide management with a metric to gauge the health of the product development process.

The following is P&W's grassroots journey to adopt elements of "Success is Assured." The change was the result of a collaborative effort between management and employees, not from a top-down initiative influenced by the continuous improvement trend *du jour*. P&W recognized that they needed to reduce rework from late learning to meet their cost and quality objectives. And they agreed that their current process had some shortcomings. It didn't hurt that they had a strong reason to change: a new revolutionary engine that had to be successful to get back into the small commercial engine market.

P&W'S Grassroots Journey to Establishing "Success is Assured"

Establishing "Success is Assured" does not follow the typical pattern of change initiatives — most of which are focused on cost reduction, process productivity, and execution time. "Success is Assured" is about the quality of decisions. On the surface, taking the time to learn first before deciding works against the cost and speed of making the decisions, even though we know intuitively that poor decisions always cost more at the end and negatively impact the schedule. Thus, changing your development process to establish "Success is Assured" is not a natural extension of any of the ongoing initiatives at P&W.

Brian and I didn't have the roadmap that this book provides. What we did have was some grassroots understanding and leadership focused on continuous improvement. This section outlines our journey, which was largely organic. We grew "Success is Assured" thinking into each of the current initiatives; it seemed like every step was another plank on the bridge — some planned, some accidental. Was every step required? Maybe not, but I think the story is important for getting the right mindset for this type of organizational change.

Why Did We Need to Change?

John P. Kotter, author of *Leading Change*, believes that instilling and maintaining a sense of urgency within a company is necessary to create lasting change. Fortunately, P&W did have a compelling sense of urgency when we started our "Success is Assured" journey: P&W's Geared Turbofan™. Its technology promised increased efficiency and reliability that allowed P&W to negotiate exclusive deals with both Bombardier and Mitsubishi Heavy Industries for their upcoming single-aisle aircraft. This was also P&W's answer to getting into the small commercial engine business. But those customers needed the new PurePower® engine to be delivered on time; the delays and cost overruns experienced on the prior PW6000 could not be tolerated.

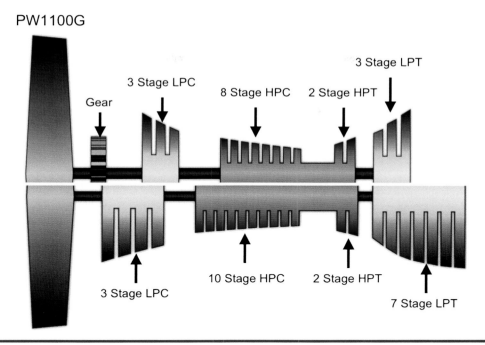

Figure 1.4   Geared Turbofan™ (top) vs. traditional design (bottom). (Image: Pratt & Whitney.)

What made the Geared Turbofan™ better than current technology? Figure 1.4 shows the difference between P&W's Geared Turbofan™ (PW1100G) engine at the top and a typical two-spool engine without such a reduction gear at the bottom.[11] Using a gear to separate the engine fan from the low-pressure compressor and turbine allows each of the modules to operate at their optimum speeds. The fan can rotate slower while the low-pressure compressor and turbine operate at high speed, increasing engine efficiency. The new engine delivers 16% less fuel consumption, 30%–50% less emissions, and 50%–70% less noise (slower rotating fan). This increased efficiency also translates to fewer engine stages and parts for lower weight and reduced maintenance costs.[12] Up to this point, only small reduction gears were used on old (World War II and before) propeller engines and small jets used on business planes. P&W needed a flawless development program to ensure customers and regulatory agencies that the bigger gear they were designing was ready for a broader customer base.

What Were Our Current Product Development Capabilities?

Given the importance of the PurePower® engine program, P&W management wanted to ensure a robust development process. Management requested a value

---

[11] P&W Geared Turbofan™ cross-sections. Online: http://www.carry-on.com.au/wp-content/uploads/2013/06/Final-Engine-Comparison-Diagram.jpg (accessed October 30, 2016).

[12] PurePower® PW1000G Engine. Online: http://www.pw.utc.com/PurePowerPW1000G_Engine#sthash.x8hbWCTA.dpuf accessed April 19, 2016).

stream mapping event on the current process. I met Brian Gracias,[13] my future partner in this journey, at the event in 2007. As members of the value stream mapping team, we were asked to identify non-value-added process steps and to create a plan to reduce the overall development time and cost without jeopardizing quality or customer satisfaction.

A few of the team members had participated in the creation of the Systems Engineering organization and the phase gate process, including the IPD value stream map shown in Figure 1.5. I don't want anybody to study or understand the details in this map; rather, I just want to illustrate the rough complexity of the process we were working with.

They suggested the team use this as a starting point for our mapping event. The map caused my eyes to glaze over and my head to hurt. How could anyone possibly value stream map this? There are so many processes intertwined, information flowing in all directions, iterations going on intentionally and unintentionally, where do you start? How long the whole process takes isn't predictable, especially when developing a complex product. There are so many people inside and outside the company involved it's hard to know what process sequence is optimum. And with the countless hand-offs between groups, how do we know when each hand-off should occur? How do you define the critical path? What are the long lead time parts that need to be defined early to ensure they are

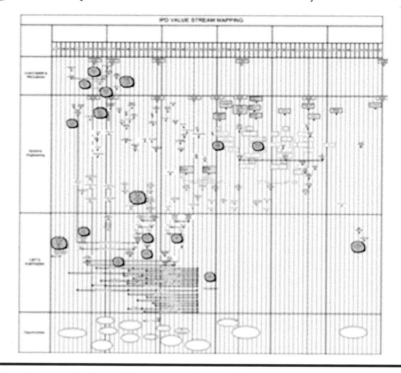

Figure 1.5 Rough complexity of the IPD value stream map. (Image: Pratt & Whitney.)

---

[13] Brian is a 32-year veteran of the jet engine business. The first half of his career was spent as a structures and design engineer. The second half focused on "continuous improvement," applying Design, Measure, Analyze, Improve, and Control (DMAIC) and Design for Six Sigma (DFSS) and Lean principles within the product development environment.

available for the testing schedule? This is primarily a chaotic learning environment — not at all conducive to a series of planned process steps.

Value stream mapping works well on reasonably linear transactional processes where hand-offs are well defined and the outcome is known. Ordering a meal at a fast food restaurant is a good example.

- Order meals from a set selection of choices
- Pay for the meal
- Prepare the meal
- Deliver the meal

The hand-offs are easily identified, options are controlled and the outcome is hopefully a good meal. There are certainly transactional processes within any development process that can be successfully mapped. However, what if you are trying to improve the quality and lead time of the overall development process? How do you know which transactional processes influence the overall development process? Companies all over the world are value stream mapping their technical processes and declaring success, but are they improving their overall development process? Who owns the product development process and how does a company with a very complex product measure success? These products can take years to develop and requirements can change in the middle of the program. Value stream mapping P&W's development process didn't make any sense to me.

Our team did try to build an orderly process. We spent several meetings discussing alternative critical paths, when hand-offs should occur, and how suppliers affect the process. Many members thought the real issue was the number of iterations allowed on a design. They believed engineers kept iterating a design to achieve perfection even after the customer requirements had been achieved.

That didn't make sense to me. In my experience design iterations were done out of necessity, not because I had extra time! Instead, iterations were the result of wrong or late input data, bad decisions, or even revised input data late in the design.

Brian also questioned the direction the team was heading. He wanted to see data on previous programs. He explored key questions, including

- What were the major loopbacks?
- Where was the most time spent?
- How many redesigns did we do?
- Were there examples of too much iteration?

The team's answers were not based on data. I found them subjective and anecdotal. Brian's questions helped move the team in the right direction, but neither of us had any idea how to get started.

During random hall talk, Brian and I discussed the event and agreed that limiting the number of iterations was not the answer. I realized from our discussions that Brian and I had similar ideas about process improvement in an engineering environment. But where would we start? I knew I had found a kindred spirit for advice and brainstorming sessions.

The team created a report with suggestions on how to reduce waste and improve quality. I was not involved in implementing the suggestions. I wanted to go back to my own group and experiment with my ideas to reduce waste and improve quality in an engineering environment. I started calling this my "equally informed people" master plan.

Brian and I kept in touch after the value stream event. We relied on *management by wandering around* (MBWA): connecting at business meetings and chance hallway meetings. We discussed our latest successes and failures with a continuous and jointly played game of "What would I do if I were in charge?"

Brian and I realized that we didn't know how to improve highly variable and iterative learning processes. Our current continuous improvement tools alone were not going to solve our problems. We needed a way to find the critical path to optimize the learning cycles that are fundamentally chaotic in nature.

How Does Organizational Complexity Affect Complex Processes?

Between 2007 and 2009 I managed the Configuration Management (CM) organization, which was part of the Systems Engineering organization. In addition to designing the engine system structure, system designers manage change requests for current production engines and development engines. The CM group manages the bills of material for the engine programs and ensures internal and regulatory agency process requirements are followed.

The change request process crosses multiple organizations, similar to a development process but smaller in scope and not as complex. Still, the change process provides insight into how organizational complexity affects a complex process and how you can minimize the impacts.

Engineers trying to execute changes to the bill of material complained that the process was too complicated and time-consuming. Completing a change involved many hand-offs and inputs from engineering, manufacturing, suppliers, and field support. Engineers wanted it simplified and many people suggested conducting a value stream mapping event to eliminate process waste.

Given my experience with complex process improvement, I wasn't enthusiastic about this suggestion. Instead, I wanted to test my "equally informed people" idea.

Several of my managers told me that we (the CM group) were getting complaints from the Component Centers. We were not processing engineering changes fast enough for a critical development program. The Component Centers wanted us to hire more people to solve the problem.

The urgency resulted because P&W promised the customer that the next round of delivered engines would contain the design improvements identified during the early development program testing. Also, the customer was holding P&W to a tight schedule. This meant my team had to process engineering changes quickly, so that manufacturing had the necessary time to incorporate the changes. Of course, no one wanted to be the cause of late engine deliveries, so everyone blamed someone else and that someone was usually CM.

To me, the problem was obvious: no one really understood the engineering change process or what his or her responsibilities were to the process. If they did, then we wouldn't have a problem. (Equally informed people seldom disagree!)

There was a comprehensive computer system in use to manage the current bill of material and pending changes. The program tracked the dates a change entered and exited a major decision gate. With this information I was able to create a simple visual chart (Figure 1.6) that illustrated where the pending engineering changes were in process (what decision point they were at) and what discipline was responsible for moving them to the next process step.

The PCE for the program loved the chart. He knew immediately where a change was in the process and who was responsible for moving the change to the next gate. He could also tell by the progress if the engineers were following his priority instructions. Prior to this chart all he heard was that CM was holding up the engineering change, which wasn't accurate or actionable.

With this simple visual chart, the PCE could share the change prioritization with everyone and eliminate guessing or miscommunication between groups. Suddenly and with little fanfare, changes moved smoothly through the process. CM didn't have to add staffing. My team also realized that the process training material wasn't written for the process users, it was written for the CM group, so we made changes so that the training addressed all the process players' needs.

So what really happened here? CM provided a simple visual model of the process to the users so they could communicate more effectively and improve the

				Group Responsible to Obtain Decision						
				Design/Suppler	Design/ Supplier	Configuration Mgmt	Manufacturing	Design/ Supplier	Design/ Supplier	Design/ Supplier
				Major Decision Points in Process						
Engine Program	Engineering Change Number	Change Description	Responsible Design/ Supplier Group	Decision Gate 1 Approval	Decision Gate 2 Approval	Decision Gate 3 Approval	Decision Gate 4 Approval	Decision Gate 5 Approval	Decision Gate 6 Approval	Final Approval Gate
YYY1	12345	XXXXX	High Compressor	Date	Date	Date				
YYY1	22345	XXXXX	High Turbine	Date	Date					
ZZZ1	5678	XXXXX	High Compressor	Date	Date	Date	Date	Date		
ZZZ1	6678	XXXXX	Externals	Date						
ZZZ1	1122	XXXXX	Fan							

Figure 1.6   Table of changes and the major decision gates. (Image: Pratt & Whitney.)

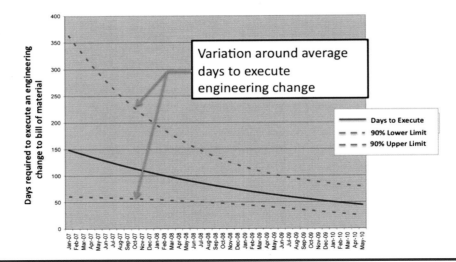

Figure 1.7   Reduction in both time and variability to execute changes. (Image: Pratt & Whitney.)

process performance. The training also helped engineers reprioritize their tasks so they synchronized with their roles and responsibilities in the change process.

Over a three-year period, I was able to demonstrate a significant reduction in the time to execute engineering changes and the variability around the process, as shown in Figure 1.7.

Figure 1.7 shows the actual times to complete engineering changes that affected a bill of material. In addition to a reduction in the average time to complete a change, the overall variation was reduced. Between 2007 and 2009, the average time to execute a change reduced 67%, while the overall variation was reduced by 80%. All this by providing visuals and training, and not adding any more staff!

By breaking the complex change management process into Important Decision points, I helped the organizations stay aligned. The alignment across the groups meant they had the data at the same time, allowing them to make decisions that improved the product and process with minimal negative impact to the other players in the process. We could have done a value stream map of the process, but I don't think all the formal requirements of the mapping event were necessary here. The focus of the effort was making the major decision point status visible and the associated process data needed to make better decisions. Equally informed people seldom disagree!

I hadn't realized this yet, but a similar approach is used in establishing "Success is Assured" when facing product development decision points. And the same "aha" moment applies.

I realized from this that in complex organizations designing complex products, the right knowledge must be available to each decision-maker at the right time. Without that, bad decisions are inevitable. And top management needs to understand that.

The Foundation for Establishing "Success is Assured"

Brian is a teacher and course developer for the continuous improvement curriculum at P&W. In one of his Lean courses he reviews Dr. Allen Ward's *Lean Product and Process Development*.[14] After studying the Toyota product development process, Allen identified their underlying principles for success:

- Set-based concurrent engineering (SBCE)
- Team of responsible experts
- Cadence, pull, and flow (CPF)
- Entrepreneurial system designer

Brian used Figure 1.8 to show the class how companies should organize their approach to product development. He thought the model could help guide and streamline our journey. After studying the model Brian and I felt comfortable that we understood and were proficient in two of the pillars:

- P&W had strong teams of responsible experts (our technical experts assigned to reinforce technical knowledge in our decentralized engineering organization).
- P&W had strong product leaders (our PCEs).

So we decided to focus on the other two pillars, both for understanding and for capability alignment.

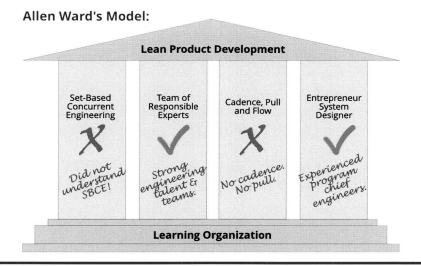

Figure 1.8   P&W's proficiency with the pillars of Lean product development.

---

[14] A.C. Ward, *Lean Product and Process Development*, Cambridge, MA: Lean Enterprise Institute, 2007.

Getting Executives to Understand Cadence, Pull, and Flow

We decided first to focus on CPF. We knew that P&W practiced CPF in local areas of engineering but not in multidiscipline complex processes. We needed managers' and senior managers' support to institute CPF, but how would we get them trained? Brian and I spent many lunch hours trying to figure out our next steps.

We needed a novel approach and we made a risky and potentially career-limiting decision: get the executives to play the Bead Game.[15] In his training sessions, Brian's students often asked if the executives had played the Bead Game. They thought leadership needed the training. "Didn't the execs realize that they drove the multitasking and constant priority changes?"

If we could get the executive team to play the game, we might get more support in applying CPF across interdisciplinary processes. Brian and I spoke with our bosses about our idea. We asked how we could hijack the engineering VP's staff meeting so his direct reports could play the game. They agreed to get the VP's permission to use his staff meeting for such a training event.

In the Bead Game, teams execute multiple process steps on multiple projects while performing simple tasks with glass beads. It simulates critical-chain thinking and demonstrates the detrimental effects on flow from multitasking, changing priorities, and driving task-based efficiencies.

On the day of the meeting, Brian and I were worried. We didn't know if the executives would be angry or happy that we sprung a game on them. And when the meeting started, their faces were not happy.

The group was broken into four teams of five people. Each team received three sets of colored beads. Written instructions were provided on how to process the beads using paper plates and spoons. Each team member was assigned a specific task in the process. The goal was to complete the processing of each of the colored bead groups as quickly as possible. To simulate multitasking, the teams were required to start work on the second and third color beads as soon as imaginary money was released.

This gave the teams about two minutes to work on one color bead group before money was released for the next color group. There were strict rules on how the team could manipulate the beads. Task completion time for each team was recorded on a flip chart.

Eventually, there was a lot of laughing and confusion during the game. No sad faces! When asked how they liked the game they said it was "confusing, chaotic, and not a lot of fun." I asked if this felt like a typical day at P&W. No consensus: some said yes, others said no. I also asked what they would do differently to help the game go faster. The answer was to use bigger spoons. Bigger spoons in the real world equates to adding more people, IT systems, machines, and so on

---

[15] Tony Rizzo invented the Bead Game in the 1990s while employed by Lucent. He later became the president of Product Development Institute.

to the task. I grimaced; that seemed to be the standard solution for every process issue.

They then repeated the game, with only one change to the process: this time they had to finish one bead color task before starting another, regardless of when the money was released. The second game went more smoothly and each team reduced their time to completion by more than half. When asked why they were more successful in the second game, they said it was because they could focus on the task to completion and they had a better idea of what they were doing after the first game. One brave executive said that the people in the room were their own worst enemy. They caused much of the chaos by interrupting their employees with panic jobs and priority changes all the time.

Good news (for us!): Brian and I weren't fired. We were actually thanked for holding the training. The executives spent time after the training brainstorming how they might apply this learning. They selected a process that they perceived was slowing down the PurePower® development program: the rotor dynamics analysis process. I was asked to apply CPF tools to the process and demonstrate lead time reductions, while maintaining or improving the quality of the results.

> You should act on any chance you can get to help senior management become involved in process improvement. They have a much bigger picture of the business and where the problem areas exist. Because they are pulled in many directions, they often don't get the chance to coalesce their thoughts and find solutions.
>
> ■ The Bead Game allowed them the time to understand the issues and a possible solution. The process the P&W executives selected was the best one to prove out CPF on complex multi-organizational processes.
> ■ Because the executives selected the pilot process, they felt ownership in the results. When I ran into roadblocks, they were vested in helping me resolve them.

Applying Cadence, Pull, and Flow to Rotor Dynamics Analysis

The executives selected the rotor dynamics analysis to prove out CPF. It is a complex process, more so than the engineering change process, and is critical to meeting customer and regulatory requirements. Excessive engine vibration can affect airplane wing performance, cause cabin shaking, and reduce engine longevity. Think of the tires on your car: when one is out of balance the whole car shakes, the ride is uncomfortable and noisy, and if not corrected, the unbalanced tire will wear out faster than the other tires, possibly causing an accident. Someone has to make sure all the parts in the engine are designed and positioned properly so the engine doesn't shake excessively and negatively affect engine and airplane performance.

At the start of an engine design, system designers produce an engine cross-section for analysis. They use low-fidelity analytical tools to position critical components to meet vibration requirements. For example:

■ Bearing locations
■ Mount locations between the engine and aircraft
■ Airfoil counts and diameters
■ Structural member locations to ensure appropriate engine stiffness

Once they are satisfied that the engine will be able to meet vibration limits and other basic customer requirements such as weight, performance, noise, and overall engine length and diameter, they send the cross-section to the responsible Engineering Component and Design groups and the Engine Dynamics and Loads (ED&L) group, who perform a higher-fidelity analysis. Rotor dynamics analysis determines if the bearing and engine mount locations are in the optimal position to meet vibration limits. ED&L also defines the center of gravity location for the major engine parts and passes this on to the part and component designers. Component designs must meet the calculated center of gravity while also meeting their ESW requirements. This analytical process iterates with increased fidelity until all requirements are achieved.

Everyone is constrained by schedules. There are schedules for the overall product development process, raw materials, manufacturing lead times, and engine assembly lead times. The design teams can't wait for late information from ED&L. Designers need to get their blueprints released so that hardware can be delivered on time for testing. When ED&L misses the scheduled milestone, designers have to use old, lower-fidelity information to design their parts. Later, if designers find that the released blueprints don't meet ED&L's final design requirements, they need to iterate on their design again. This ripples downstream to tooling changes, material expediting, part rework, and so on, all costly and ripe for quality and delivery issues. Sometimes, the corrections cannot be made until after the engines are delivered to the customer, causing engine removals at inconvenient times and unhappy customers.

As I dug into the ED&L process I discovered similarities between this process and the engineering change process.

■ People didn't understand their role in the process.
■ Important decision points weren't broadly communicated.
■ Inputs to the process were late, out of sequence, or just missing.

I worked with ED&L to create a very simple database to track all the current tasks and pending tasks based on key decision points in the process. Once we had a simple comprehensive method to track tasks, I asked the PCE for the PurePower® engine family to prioritize the work. He was surprised at some of the task and model change requests. This explained why the PCEs complained about the accuracy of the rotor dynamic analysis results: the problem wasn't the quality of the analysis so much as it was that the cross-section originally approved for analysis wasn't what was finally analyzed.

ED&L applied the new task rules to a PurePower® vibration analysis. With the new prioritization process and a better understanding of input accuracy and timing, we realized significant improvements to the lead time and results accuracy. Brian and I had our first real success.

The PurePower® analysis was completed two weeks after the original promise date. While that may not sound like improvement, consider that the delivery of results was often two to three times that and sometimes even more. The development program can handle a one- or two-week delay, but delays greater than a month start to impact the development schedule and the accuracy of the individual part models used in the designs.

The ED&L team was excited with the results and wanted to continue to improve their process. They spoke positively of the success and the executives started taking notice.

Understanding the Last Pillar: Set-Based Concurrent Engineering

While I was working with the Rotor Dynamics group, Brian started investigating the last pillar in Allen Ward's model: SBCE. We needed help to understand this last pillar. Brian looked outside aerospace and other UTC divisions to benchmark other companies successfully using SBCE.

Brian found the Lean Product and Process Development Exchange (LPPDE). The website described the conference in the following way:

> *The Lean Product and Process Development Exchange, Inc. is a non-profit organization created to foster opportunities to grow and share the knowledge, expertise and experiences that help organizations use Lean product development to dramatically improve product development performance.*[16]

They were hosting a conference in San Diego, California in March 2011. On the board of directors was Michael Kennedy, who Brian met many years ago while working at Honeywell. Michael had written a couple of books on Lean product development (LPD) and is a sought-after consultant by many major U.S. and foreign companies. Brian thought we should attend the conference and see if we could get Michael to help us on our journey.

The presentations were by companies on their LPD journeys, sharing what worked and what didn't work. Knowing others were on the same journey and shared similar difficulties was helpful. Brian and I met with Michael Kennedy

---

[16] Lean Product and Process Development Exchange. Online: http://www.lppde.org/About-LPPDE (accessed 10/30/2016).

and another consultant, Ron Marsiglio.[17] Ron and Mike often worked together. We liked what we heard, so we invited them for a visit to P&W (the start of an ongoing mentoring relationship).

They visited us in late summer 2011. We reviewed our current development process and then Ron and Mike gave examples of SBCE establishing "Success is Assured" and the benefits. They also said that almost every company they worked with used a phase gate process and requirements flowdown to manage their product development process, and they all experienced loopbacks, resulting in schedule and cost overruns.

So what's different between an SBCE process establishing that "Success is Assured" and the typical point-based requirements flowdown process? Figures 1.9 and 1.10 contrast the philosophical differences.

Requirements are parceled out to each subsystem. Teams work to meet their assigned requirements (task based). The PCEs pull the teams together several times before a major phase gate review to assess progress. At the end of the planning or preliminary design phase a configuration is selected for detailed design and testing. Risk mitigation plans are created and tracked for those aspects of the configuration that do not meet requirements. Configuration testing is the vehicle used to validate that requirements are met or identify requirement shortfalls. Given the learning occurring late in the process, when little can be done to recover from surprises, the time frame of the process and the quality of the results both tend to be unstable.

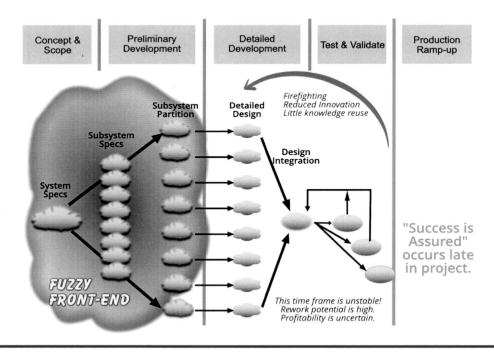

Figure 1.9   Point-based requirements flowdown process.

---

[17] Ron Marsiglio is the owner of Knowledge/PD, LLC, a consulting firm focused on knowledge-driven product development.

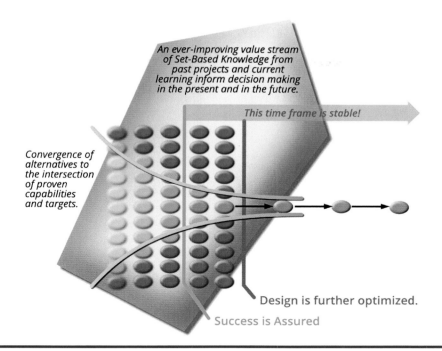

Figure 1.10 Set-based concurrent engineering (SBCE) process.

The Systems Engineering organization and the component design teams work collaboratively to assess many configurations. Requirement targets are set instead of hard goals. Teams use specific enabling tools (to be described in Chapter 3) to identify knowledge gaps and explore the available design space. Sets of configurations are assessed, and knowledge gaps are identified and closed before decisions are made. The teams wait as long as possible to pick the configuration for detailed design. This method minimized the learning during testing, which reduces lead times and costs.

Note that "sets of configurations" does not mean discrete options. Rather, they are continuous sets that define the design space of viable solutions, as shown in Figure 1.11. Any combination of *X*- and *Y*-values in the unshaded region of that

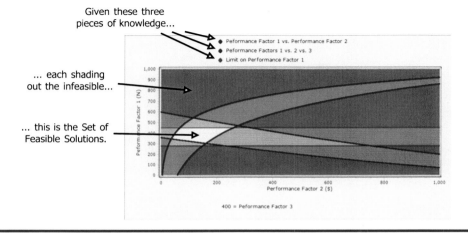

Figure 1.11 Generic depiction of the design space of viable solutions.

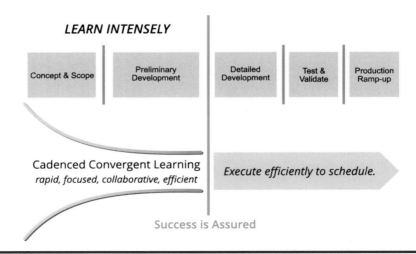

Figure 1.12   Impact of SBCE on the gated development process.

chart is feasible. So, there are an infinite number of options that work. Evaluating each individually is unnecessary if you can instead evaluate where the boundaries of that region are. Each piece of knowledge can introduce additional boundaries to that design space.

By establishing that "Success is Assured," SBCE arrives at a configuration with less rework and fewer issues in the field. This results in lower overall operating costs, faster product development, and happier customers. We will go into more detail in Chapter 3.

Figure 1.12 depicts the impact of SBCE on the gated development process. You don't select the final product design until you have developed knowledge that shows you've selected the best options and you are sure your development program will produce a successful product with minimal or no costly loopbacks. The learning is pulled up front into a convergent decision-making process, enabling the more expensive back end to flow smoothly without rework. Each new piece of knowledge that introduces an additional limit reduces the remaining design space, thereby converging the decisions a bit more.

Can You Apply Set-Based Concurrent Engineering Practices to Complex Systems? Establishing Relevance to Jet Engines

Brian and I took Ron and Mike on a tour of our P&W engine maintenance–training center. We wanted to make sure they understood the complexity of jet engines. The training center has several engines on display. This facility is used to train airline mechanics and overhaul shops on engine maintenance methods and practices. We showed them the outsides of different engine models. The engines are completely covered with tubing, valves, sensors, and other devices that ensure the engines perform to requirements in all operating environments.

The expressions on their faces were priceless. They were gaining much greater appreciation for the level of complexity that would have to be managed in order to establish "Success is Assured." Complexity adds many interdependencies to the overall system. Engineers often get lost in the interdependencies and miss important decisions and risks. P&W engineers call these the *unknown unknowns*. History has shown the best way to find these unknowns and address them is through extensive testing of the system and major subsystems. Unfortunately, finding most of the unknowns in the testing phase creates a cascading flow of design rework.

The key question still remained: how do you apply set-based strategies to a complex product or how do you find the unknown unknowns before testing?

> I learned that Allen Ward's LPD model provided us some great insight into what was needed to establish "Success is Assured" in our system. I already understood the importance of the system designer and the expert workforce. Now I needed to add CPF to efficiently pull the learning to close knowledge gaps and SBCE to keep multiple options open until knowledge gaps were closed. Otherwise, bad decisions would be made and cause rework that would be very costly to the product development program.
>
> The most important learning was that the Ward model was a system, not a pieced-together set of principles; all four pillars are required and they all have to work together in order to establish "Success is Assured" prior to decision-making. (The details of those interdependencies will be covered in Chapter 3.)

Refining Our Phase Gate System with This New Thinking

As Brian and I were plotting our next steps, we found an unexpected partner in our journey: Pratt & Whitney Canada (P&WC). P&WC, based in Longueuil, Quebec, is a sister company to P&W in the UTC Aerospace division. The company's products power business and regional aircraft and helicopters (more than 60,000 engines in over 200 countries and territories).

P&W and P&WC were involved in restructuring efforts to improve synergies between the two companies. Goals included reducing the cost and lead times of product development and field redesigns. Each P&W engineering discipline was asked to work with their P&WC counterpart to share best practices and find synergies. The leader of the Systems Engineering organization for P&W was aware of our LPD journey and had met with Ron and Mike during their visit. After a few meetings with P&WC, she contacted Brian and I thinking we would be interested in P&WC's product development process. She thought their process had aspects of "Success is Assured" that might help P&W move in that direction.

P&W's phase gate processes, like all phase gate processes, were very linear. Integrated Product Teams (IPTs) reviewed their designs with the local Technical Fellows. If the review was passed, the results flowed to the Module review. The

Module review was performed with the Technical Fellows and Module leadership. Results were fed to the PCE for inclusion in the system-level review. The process contained many redundancies.

P&WC also used the phase gate process, but they adopted a review process that was more parallel than series. They held technical reviews between system reviews that included representation from all the company disciplines and sometimes suppliers. These meetings could last for three days if necessary. All the technical experts and product designers and systems engineers were in attendance. This allowed collaboration to quickly identify changes to one module that would adversely affect another module or the overall system. Actions to correct the issues began immediately. All the necessary experts were in the room to approve the change.

P&WC held as many technical reviews as needed between systems reviews. And they all agreed on the risk plans, resulting design actions, and the presentation contents to be given to the System Chief Engineers at the system review gate. These meetings sounded a lot like Ron and Mike's description of how other companies converge on product configurations before detailed design. They called them Integrating Events (to be described in Chapter 3).

Our Systems Engineering organization recommended adopting P&WC's process, and our management agreed. Brian and I asked that the technical reviews be called Integrating Events. We wanted to start using the "Success is Assured" process names wherever possible. Part of change management is familiarizing people with terms. We believed the sooner we started using "Success is Assured" terminology the better.

> This was a big step: Brian and I had found a way to integrate "Success is Assured" behaviors into our phase gate process. Both processes have their place in the development and execution of new products: set-based Integrating Events to pull focused learning and phase gates for efficient task execution. Our challenge: Brian and I still had to figure out how to make the overall system as effective as possible.

Implementing a System for Establishing "Success is Assured"

I was promoted to Senior Fellow Discipline Lead (SFDL) of Systems Engineering at the end of 2011. This put me smack in the middle of the most esteemed technical leaders at P&W. Each major discipline at P&W has a Senior Fellow who is responsible for improving his or her discipline's analytical capabilities and staying current with the latest external information and technology in their discipline. SFDLs are also responsible for ensuring their discipline had the breadth and depth of knowledge needed to continue designing and manufacturing the best jet engines. SFDLs participated in the phase gate design reviews and now were required participants at the Integrating Events.

After several months with this group, I made the connection that they would be the right group to move engineering toward establishing "Success is Assured." Many of the SFDLs were heavily involved in the PurePower® engine development process. They often commented on how they would like to change things in the process. I liked some of their ideas and used our discussions to introduce the principles of "Success is Assured."

Near the end of 2012, the PW1500G[18] engine certification effort was winding down. Brian and I decided this was the right time to get the SFDLs together with Ron and Mike for a two-day off-site. Getting all the SFDLs in a room at the same time would be difficult and keeping them there all day was impossible, but I moved forward anyway. If they stayed and engaged with Ron and Mike, I had the right group. To my surprise, most of them stayed for the whole first day and they all showed up the next morning. They immediately grasped "Success is Assured" and its benefits. They all believed that many of our current challenges were the result of rushed decisions early in the development process.

At the end of the second day, the SFDLs suggested we get the engineering executives in a room to share the same information they just heard. They thought the time was right for this. P&W was about to complete the PW1500G engine certification process, so loopback memories were fresh in everyone's mind. Brian and I liked the idea and put a plan together to make it happen.

Each SFDL met with their executive to fill them in on our recent meeting and ask if they would participate in an off-site to listen to our ideas. Reception was positive, so I set the off-site for mid-December 2012. All the executives agreed to give us 1 day. The SFDLs wanted Ron and Mike to give the same presentations to the executives that they had received, and each SFDL wanted to give a quick presentation on their thoughts and ideas. Brian and I were to facilitate the meeting. Our goal was to develop a list of action items that would support establishing "Success is Assured" in our product development process.

I frequently discussed the ups and downs of my LPD journey with my boss. He was very supportive and gave me free reign to pursue the journey. He also gave me great advice on how to handle some of the more difficult people I encountered. But the best advice he gave me was to invite Chris Kmetz, the PCE for the PW1500G engine program, to the off-site. He knew Chris was an early adopter and he had the respect of all the engineering executives. If Chris liked what he heard at the meeting he might be able to influence the others.

Brian and I arrived at the off-site early to make sure the room was set up correctly and the lunches were ordered and delivered on time. We paced in the room for 20 minutes waiting for someone to arrive. Fortunately, all the executives were on time and the meeting started without any hitches. Ron and Mike

---

[18] The PW1500G is the engine used on the Bombardier C series aircraft. P&W website, Commercial engines, PW1500G product card. Online: http://www.pratt-whitney.com/Content/Press_Kits/pdf/ce_pw1500g_pCard.pdf (accessed February 9, 2017).

were the initial speakers. I was pleasantly surprised that when the executives challenged Ron or Mike, the SFDLs spoke up to answer the challenge. They really supported the initiative. Then several of the SFDLs spoke. They reviewed issues in the current development process and gave examples of how spending more time in the early phases establishing "Success is Assured" could have minimized these issues. The exchanges were interesting and did uncover some resistance.

■ Everyone agreed that we often miss development milestones due to aggressive schedules.
■ Missing schedule requirements is to be expected with a complex product because it's impossible to identify all the risks involved or all the possible setbacks.
■ Studying several configurations before we move to preliminary design is already part of our product development process. How is "Success is Assured" different?
■ Informing leadership with our phase gate reviews works well. Why would we stop them?
■ Including knowledge curves in our ESW documents is part of the current process. Why would we want to change our process?

These comments didn't surprise me. Why would they think any differently? Brian and I weren't in a position to show them a comparison of our product development process to "Success is Assured"; we were still in the discovery process. We had a lot more work to do before we could adequately challenge the executives' view of our development process.

At that point, Chris Kmetz was a great help. He agreed with the loopbacks identified by the SFDLs and offered more examples. Chris said that schedule demands often drove his team to make quick decisions. And sometimes he didn't realize what he didn't know, meaning he wouldn't know that he needed more time in the decision-making process. He thought modifying the phase gate process to add Integrating Events was a great start because it is part of the early discovery of unknown unknowns. At least everyone in the Integrating Event would be involved in the decision process to understand the knowledge gaps and associated risks. And if there was more P&W could do to help the design process, Chris was willing to give it a try. Brian and I were in the back of the room sighing with relief. Maybe it was time to get the team thinking of next steps.

Brian and I asked everyone to break up into teams of five and list what we should do next. These were the results:

■ Develop a P&W value stream vision for establishing "Success is Assured"
■ Perform an after-action review on the PW1500G engine development program. (An after-action review is a structured event to identify what didn't go

as planned and how actions could be improved to avoid similar problems in the next event.)
- Identify the key drivers of loopbacks
- Determine root causes and corrective actions
- Use of K-Briefs (A3s) to capture the learning

I was happy with the outcome of this meeting. The executives gained a better understanding of "Success is Assured." I did not expect them to buy into changing our current process immediately. My objective was an honest discussion about the positives and negatives of our current development process, and we did that. Getting Chris's support legitimized the discussion and allowed people to start accepting the need for change. I learned in our benchmarking trip that involving a well-respected leader is necessary to demonstrate positive results using "Success is Assured." Chris agreed to sponsor the after-action review. We were on our way!

Action 1: Develop a P&W Value Stream Vision for Establishing "Success is Assured"

In January 2013, the SFDLs met with several executives to work on P&W's "Success is Assured" vision. The model in Figure 1.13 illustrates the results.

I need to point out that in 2014 P&W believed that "Success is Assured" must be achieved before *leaving* detailed design. You will notice throughout the rest of the book we state that "Success is Assured" must be established before critical

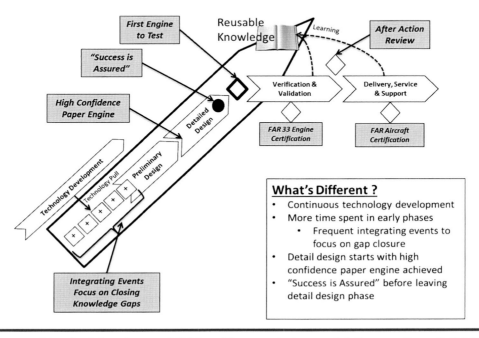

Figure 1.13   P&W's vision for establishing "Success is Assured." (Image: Pratt & Whitney.)

decisions are committed to, and a great many critical decisions become much more committed as you *enter* detailed design. I now believe that "Success is Assured" must be established before *starting* detailed design. This will further reduce the loopbacks and reinforce the behaviors needed to drive the decision process earlier in the development program.

Actually, the mapping session provided something equally important to P&W's "Success is Assured" vision. The SFDLs had an "aha" moment as we were discussing how we should structure the Integrating Events. We all knew that late or incorrect system configuration decisions early in the development program lead to costly loopbacks late in the process. These loopbacks cause redesign ripples across the engine. So, what if specific system-level decisions were tied to each Integrating Event? This would help focus the resources across engineering and more effectively close knowledge gaps.

For example, system-level decisions might include

- Structural case configuration and position in engine
- Bearing positions
- Engine Mount position
- Number of airfoil stages
- Fan diameter

Large structural case castings and forgings have very long lead times. Making raw material shape decisions early in the process is critical to maintaining schedules. This can force us to make decisions without all the required knowledge, such as exact bearing and engine mount positions. If the bearing and mount locations change later in the development process, we could face a lot of rework.

P&W's current phase gates were schedule driven and each discipline had prescriptive ESW that had to be completed before a phase gate review. This approach didn't encourage collaboration across disciplines. Inputs from group to group weren't well coordinated, as we learned from the rotor dynamics example.

If early design decisions were sequenced by a series of Integrating Events that not only finalized the decisions, but also pulled the rapid learning to close the knowledge gaps, then alignment would improve across decisions. This in turn would improve decision quality.

Systems engineers can't make these early system-level decisions alone. They need all the disciplines to focus on the early systems decisions to eliminate or significantly reduce later loopbacks in the development program. Some experienced PCEs and systems engineers may know the best sequence to make decisions based on their years of experience. However, newer program chief or systems engineers will not have yet gained enough knowledge. An ordered decision list would be very helpful, and the Integrating Events will provide the mechanism to ensure all the right disciplines are participating in the decisions.

The meeting closed with a roadmap for establishing "Success is Assured," including a preliminary list of system decisions required at each Integrating Event and a plan to continue looking into system-level decisions.

> The ordered list of system-level decisions was a great idea. Remember the struggle to identify the critical path in the product development value stream mapping event. These decisions were a first step at finding the critical path. The ordered decision list would provide a way to align all the organizations more effectively. In later chapters you will see this is a critical enabler to "Success is Assured."
>
> Some food for thought:
>
> I frequently heard people say, "There is no way you can eliminate all the loopbacks in a jet engine development program." This is true, but my response was, "I don't expect to eliminate all the loopbacks, but imagine a world where we eliminated half the loopbacks; isn't that a great start? And better yet, will the knowledge we built in doing so help us deal with the other half when they occur?" This helped others put "Success is Assured" into perspective. It's not meant to cure all ills, but it will make the overall development program run more smoothly, saving money, resources, and time. That vision made people smile. And of course, continuous improvement is just that.

Action 2: The PW1500G Engine After-Action Review

In March 2013, Chris Kmetz hosted an after-action review on the PW1500G engine development program. Approximately 80 people involved in the program from disciplines across the company (including both systems and module/component teams) attended. Chris told the group he expected them to identify the issues with the biggest impact on program cost, quality, and lead time. At the end of the meeting, the group would select the top-five issues that if addressed would benefit the next PurePower® engine development program. Later, teams would be formed to identify the root cause and corrective actions of the selected issues.

The group received a condensed version of the presentation given to the SFDLs and the engineering executives. After that, the group broke up into teams with like experiences and interests. Chris asked the eight teams to identify the issues they thought had the biggest impacts on the current development program, to rank the list in order of most to least important, and to provide data as backup. The teams had three hours to work.

I was pleased to see that their findings matched well with what Brian and I, the SFDLs, and Ron and Michael predicted. Here are a few of their comments:

- "Early architectural decisions were made without the right information and without the right people."

- "Our phase gate reviews were more task and schedule based than knowledge based."
- "Better alignment of the disciplines in the product development process would eliminate rework. Creating an organized decision process is a great start."

The teams used nominal group technique (everyone voted on the selected items, items that received the most votes were selected) to select the top-five issues they believed would benefit future PurePower® engine development programs. Chris reviewed these issues and helped select the team leaders. The next step was to train the teams.

Michael Kennedy's company Targeted Convergence provided Robust Learning training to the teams. The Robust Learning training provided tools and methods to make good decisions and create knowledge that is relevant and reusable. Tenets behind Robust Learning include

- Make knowledge visual
- Work collaboratively
- Understand design limits
- Identify knowledge gaps

Teams learned how to use Causal Mapping[19] to bring all the elements of Robust Learning together. The mapping process makes knowledge gaps visible and shows how knowledge or lack of knowledge affects decisions and customer interests. The teams gave positive feedback on the training and agreed to use what they learned when working on their after-action review teams.

Follow-On Action: Testing Causal Mapping as an Enabler for Better Decision-Making

I thought that Causal Mapping would be a good tool for systems engineers. Other disciplines have ESW that provides information such as ordered steps to design a component, how to perform the required analysis, and accepted material and design limits. Systems engineering is about integrating the overall system to ensure customer requirements are met. How do you write standard work for that? Perhaps Causal Maps could be our method for integrating components and selecting configurations. I decided to tap one of the five teams from the after-action review to test my theory. I selected a team that was dealing with an early system design challenge. Three of the team members were available to meet in my office for the mapping session: the system designer, the compressor design fellow, and the structural engineer.

---

[19] The Causal Map is a simple visual model that enables the learning required for complex engineering decision-making. Some benefits include: identifying the decisions to be made and root causes to be resolved, defining the interactions of those decisions and the required trade-offs, making the knowledge gaps visible identifying the priorities required to make the decisions. Causal Mapping will be reviewed in detail in Chapter 4 and Part II.

The team started by defining the problem statement, which they struggled to agree on. This didn't surprise me, considering all the communication issues I dealt with in CM and Rotor Dynamics. How can you collaboratively solve a problem if you don't agree on the problem! Once the team agreed on a problem statement, I asked them to identify important customer interests. The team identified the critical "must achieve" customer interests to ensure a viable product. It's not practical to use all the customer interests involved in jet engine design; teams will spiral out of control if they attempt to meet them all at once.

Customer interest selection resulted in another long discussion, again very telling. How can a team find a solution if they don't apply the same level of importance to customer interests? Next, I asked the systems engineer how he would start connecting the problem statement to the customer interests. He described the decisions he needed to make and the required input for the decisions.

He described why he selected the original fan and intermediate structural case configuration and axial spacing.

- The Acoustical group defined the acoustical treatment area needed in the cases to meet noise requirements.
- The Aerodynamics group defined the minimum fan tip clearances to meet fan stability requirements.

These translated into case size stiffness requirements. The compressor design fellow had a different perspective on case design. She had to ensure the structural integrity of the interfaces between the fan and compressor cases, which limited the acoustical treatment area. Aha, a problem! The system requirements conflicted with the module requirements. Not a big surprise. Anyone with experience designing complex systems sees these kinds of conflicts daily. The tools and methods used to establish "Success is Assured" give decision-makers a way to find these conflicts early and work out solutions. Figure 1.14 is a picture of the

Figure 1.14   P&W's first attempt at Causal Mapping. (Image: Pratt & Whitney.)

map we created. The problem statement is to the far left. (We needed four large sticky notes to define the problem!) The orange sticky notes on each side of the chart are the customer interests; the yellow notes are the decisions in order from left to right, and the remainder (light green) represent the information needed and from what groups. This map supported my vision of a process to help system engineers monitor decisions and associated inputs.

Management needs to allow decision-makers the time to find the conflicting requirements, assumptions, and knowledge gaps in order to resolve them before starting detailed design. Having the directive to establish that "Success is Assured" and the tools to do that but not being given the time to use them gains nothing other than frustration.

I considered the review a success, even though we didn't get to map the causal relationships needed for each decision. The team thought the mapping event was powerful and agreed that they would need to allow time early in the development process for the tool to be effective. Now I needed to run more mapping events and compile some successes.

I had enough evidence to be convinced that P&W should develop Causal Maps for our products. Causal Mapping describes how one decision affects another, like a technical transfer function: $y=f(x)$. Having this information for each system configuration decision would improve our product development process dramatically. (Causal Mapping is the topic of Chapter 4.)

We Continue to Test Causal Mapping

Brian and I attempted a Causal Map with turbine design experts. A jet engine turbine module is a popular improvement target. The turbine sits behind the combustor and experiences the hottest temperatures in the engine. If I could get the turbine experts to identify knowledge gaps that led to higher temperature capabilities, then I should be able to generate some interest in Causal Mapping.

The team agreed on a problem statement:

> *Design a high-pressure turbine that provides a 10% improvement in performance and durability.*

They struggled to reach a consensus on customer interests and their measurement units. This tested my facilitation skills. After about an hour, I realized that I had taken on too much. The map was looking like a spaghetti chart. I stopped the mapping and asked the team to provide feedback on the process.

- They liked that the maps provided visual information on the key decisions and cause-and-effect knowledge that was required to make those decisions.

- Any maps successfully created should be linked to our ESW. This will be a great visual to help future programs understand how to go about designing a turbine.
- They agreed I should test the mapping out on a smaller problem before I jumped to mapping a turbine module design.

Finally, the Chance to Integrate Causal Mapping into an Existing Program

While Brian and I were experimenting with Causal Maps, my management changed. My boss announced his retirement. I was happy for him but not for me. I had spent at least five years convincing him to try some non-intuitive Lean methods. There were many successes, so he trusted me and gave me free reign. Now I would have to "train" someone else. I was pleasantly surprised and relieved when Chris Kmetz was named as his replacement. I wouldn't be starting from scratch. Chris was familiar with what we were doing and was very supportive and engaged.

As the head of the Systems Engineering organization, Chris Kmetz chairs all the System Engineering gate reviews for development programs. All the system group chiefs are required attendees. They act as the non-advocate reviewers for the program. The chief engineer for the program leads the review; he or she invites those people needed to support the findings and answer the Systems Chiefs' questions. I also attended these reviews. I hoped that by educating Chris on Causal Maps he might request one to help with risk management or requirement misses during a review. This would give me a chance to test Causal Mapping on real problems and gain support for change.

I was attending a system gate review in late 2013. The program was looking to move from the preliminary design phase to the detailed design phase. The review started out as usual. The PCE reviewed the basics:

- Who is the customer?
- What is the schedule time line?
- What is the budget verses current spend and actual staffing verses planned staffing?
- What is the engine requirements status?

The PCE indicated that they were struggling to meet weight requirements. Chris stopped the meeting. He turned to me and asked me to schedule a Causal Mapping event to identify ways to meet the weight goals. He wasn't convinced that all options had been considered. He made this an action item to pass the review, meaning he would need to see the results of the Causal Map before the program received a formal pass to move to the next phase.

I wanted to jump out of my chair and do a victory dance but decided that wouldn't be professional. I finally got what I wanted: a program that needed a

Causal Map. Until then, Brian and I had been begging people to do Causal Maps; now we were getting pulled.

Brian and I met with the PCE later that week to give him a crash course in Causal Mapping and K-Briefs (Knowledge Briefs; A3s). We also gave him an overview of our journey to date and why we believed that this effort would help his program. I asked him to provide me with a list of the people who should attend. At the Causal Mapping event, we needed all the people involved or affected by configuration changes to reduce weight.

I started the event with an overview of the process and what I wanted to achieve. Then I moved to the problem statement. The problem was more complex than just meeting engine weight requirements. P&W was using an existing engine configuration in a new application. The customer needed more thrust capability than the core engine currently provided. To complicate things, P&W was competing with an engine company whose engine was already on the airframe. Our market position was to provide the same or more thrust than the current engine with a 15% performance improvement. We hoped this would make converting to our engine the customers' best option. The team's current solution to the thrust and performance target was to add a compressor stage, which meant adding weight. After some discussion, the team agreed on the following problem statement:

> *Meet the customer thrust requirements while achieving all the necessary customer interests.*

The team agreed on the targets for the key customer interests to ensure the customer would select our engine. This was another interesting but time-consuming discussion with the experts. Perseverance and patience are a necessary part of achieving equally informed people so they can agree! We finally came to consensus and listed the top-six requirements to achieve success. I posted the customer interest target values and the units of measure on the right-hand side of the map (Figure 1.15).

The team said that the key information required to start the process was the turbine inlet temperature and pressure. I listed the information and decisions needed. The team quickly identified that not all the information was available; they were working on assumptions and inaccurate data to do their ESW — possible loopbacks waiting to happen! The PCE tried to push back on this but finally admitted he didn't appreciate the importance of the data to the analysis. He received an action item to meet with senior management and nail down some decisions so the team could have all the right information.

The team continued to list inputs and decisions needed to achieve the customer interests. I mapped their decisions with sticky notes. Next I asked for possible solutions to achieve all the targeted customer interests. The team identified solution B. I can't go into technical details, but I can say that this solution identified changes to the first few existing rows of airfoils in the compressor instead of adding an additional row. The Aerodynamics group's analysis indicated that

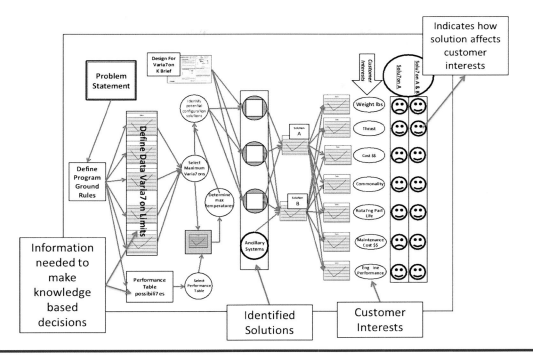

Figure 1.15   Rough layout of the Causal Map addressing engine weight concerns. (Image: Pratt & Whitney.)

the revised airfoils wouldn't meet design limits at the higher thrust levels and rejected the change.

I asked, "How extensive was the analysis?" They only analyzed two different airfoil configurations; this was a point-based approach.

Getting Set-Based Concurrent Engineering on the Table

I was hoping this would be their answer. P&W has practiced a subset of SBCE for years; they call it Design for Variation (DFV). Groups that have statistical expertise and are computer savvy tended to do more DFV than groups without those capabilities. As the Systems Engineering Discipline Chief, I worked with groups looking to reduce design time and redesigns using DFV. Roadblocks to implementing DFV include older computer analytical tools that are not DFV friendly and a lack of statistical knowledge to set up calculating methods able to compare several variables at once and map the results. I asked the team if they thought solution B may actually work if they could run more than two configurations in the analysis. The answer was yes, but their analytical tool couldn't handle several configurations.

Before the meeting I suspected DFV wasn't being used, so I had invited some of the P&W statisticians I worked with to attend the mapping event. I introduced them to the Compressor Aero group and asked that they meet to see what could be done quickly. The meeting ended with a Causal Map containing several knowledge gaps and action items to close those gaps. The team agreed to meet in a week to assess progress. The feedback on the Causal Mapping process was

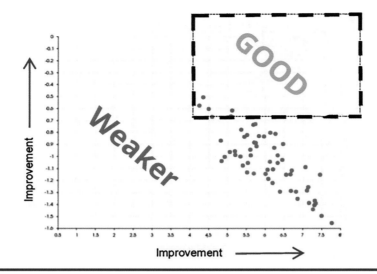

Figure 1.16   Set of alternative solutions evaluated. (Image: Pratt & Whitney.)

positive. Everyone agreed that the map solicited new and important discussions and the visual nature of the map was helpful. The team also said the event gave them a better appreciation of how they all interact together to design parts and components. This was another incremental success toward establishing "Success is Assured."

After some tweaks, the modified compressor analytical tool was up and running. They ran 10,000 configurations varying several airfoil parameters and then calculated the impact on the selected customer interests. The results are shown in Figure 1.16. The graph identifies the airfoil parameters that have a 99% chance of meeting all 18 ESW aero design requirements. The parameters I hoped would eliminate the need for an additional compressor stage are outside the good area, but there were several parameters within the good zone. This is a great example of set-based engineering. The team varied all the parameters that affect airfoil performance and compared the results to the customer requirements. They didn't "guess" which parameters would work; rather, they took the extra time to look at all the parameters to be sure they had the best solution.

Now the team knew for sure that removing a compressor airfoil stage was not a valid solution. The added compressor stage was looking like the best option. A K-Brief was created to document the assumptions and statistical methods used along with the results. Even though the outcome didn't solve our problem, the mapping session was successful.

Discussions Around Visual Models Help Pull in the Required Expertise

The team did find a way to meet the weight and thrust requirements even with an added compressor stage. The solution speaks to the power of collaborative learning and the tools that support it.

The air systems engineer supporting the program found out about the Causal Mapping event and asked why he wasn't invited. He told the PCE that he had a solution to the weight issue that wouldn't affect the other requirements, but he wasn't given the opportunity to review it with the team. Wow! Apparently, the new airframe didn't require all the air system controls that were currently on the engine. Different air system requirements meant some of the externally mounted controls and associated tubing could be removed, reducing the engine weight. Removing the controls would reduce the engine weight more than adding a compressor stage. Wow again! This was great news and now there was a weight margin going into detailed design!

So how did this happen? I asked the PCE why his team got so far without knowing about the air system differences. He admitted that he wasn't very familiar with engine air systems. It never occurred to him to include them in a thrust and weight issue discussion. The team is not co-located; had someone not thought to invite the air systems engineer, he might never have known about the issue until it is was too late. The PCE also said that he thought air systems design could be handled later in the development process. This was valuable information. Air systems design has major impacts on the overall jet engine system. I could see a strong argument for having the Integrating Events follow a list of sequenced system decisions, supported by a list of disciplines that must be at the event, as identified by the cause-and-effect relationships made visible in the Causal Maps.

This example showed many positive aspects of establishing "Success is Assured" and how it is an improvement over our current product development process.

P&W gained a lot of documented knowledge that will be useful to other teams addressing similar future compressor design challenges.

- A Causal Map was created as part of the process, it wasn't something created after the fact for documentation.
  - This tool facilitated collaboration across many different disciplines.
  - It will be a guide for future teams, showing them what customer requirements we were aiming for, what solutions were investigated, and the results.
  - The map also shows the decisions that need to be made and the disciplines needed to provide the required knowledge.
- There's a K-Brief showing the SBCE results; future teams won't have to recreate this. If customer requirements change for future products, the team can use this to assess what they need to do to meet the new requirements. They may determine that they already meet the new requirements, which will reduce their workload and keep costs down.

Leadership support is critical to establishing "Success is Assured." Teams must be allowed the time to explore the design space and collaborate with their discipline peers. Employees will find it difficult to ignore task completion and schedule requirements unless someone at a higher level supports what they are doing.

You may be wondering what happened to the after-action review teams that were going to identify root causes and corrective actions for the identified issues. I didn't get as far with the teams as I had hoped. It was very hard to get the teams to break away from their day jobs to work with us. They were very busy addressing issues identified during product testing. This is of course what happens when new methods are advocated, but existing workloads remain. Also, we didn't have the roadmap for implementing "Success is Assured" that this book supplies. That's one of the reasons for this book. Based on P&W's and other companies' experiences trying to establish "Success is Assured," I realized that just telling people to make good decisions and use SBCE wasn't enough. In Chapter 6, we provide advice on how to get started with "Success is Assured" when you don't have the luxury of starting your product development using "Success is Assured."

Where to Next: The Journey Must Continue

My LPD journey ended soon after we successfully demonstrated SBCE. I retired in October 2014 after 35 years at P&W. Why, you ask, would I retire in the middle of all this? Retiring early was always part of my career plan. If my presence was needed to make "Success is Assured" work, then Brian and I hadn't made any real progress.

What are P&W's next steps? They are using K-Briefs to collaboratively solve problems and document knowledge. They are also working on their knowledge capture and reuse process to improve the relevance of the search and retrieval capabilities. The new tool will add knowledge gaps and design space trade studies to the search data to improve relevancy. I can't speak to how far the culture has moved toward embracing "Success is Assured," but the use of Integrating Events in new technology and product development programs will help. Knowledge gaps will be identified sooner and the discussions of how a system decision affects other disciplines' standard work now have a venue.

Here are my recommendations from this grassroots journey. They might come in handy if you are planning a similar journey.

- Find a partner.
  - You need someone to bounce ideas off of and to listen to you when you get frustrated.
  - Each of you will have different professional experiences that help you create more robust plans and presentations to sell your strategy.
  - Going into meetings to discuss Lean with skeptical executives is easier when you have a partner to back you up.
- Find an early adopting leader to help you on your journey.
- Be tenacious.

- If you aren't successful with one group, don't give up: try another group. Make it your goal to turn their negative views on LPD into positive views.
- Continue learning; find out how other companies solve their development problems, read books, find as many resources as you can, and use them.

■ Create a plan for change management (and be sure to include your leadership).

- We used Ward's development model to compare and contrast and guide our learning journey. Without that, our journey would have taken much longer.
- Don't try to completely throw out the existing processes; there were good reasons they were created. Rather, find ways to align their goals and get the new ways integrated into the standard daily work.
- Also, if you tell busy people they must totally change the way they are doing their jobs, they will tune you out. You need to get their attention by including them in small successes with the new tools.

■ Try something — and then try something else if it doesn't work.

- We like to say, "Act your way into Lean thinking." If we had waited for permission to do the things we did, then we never would have gotten started!
- Leadership accepting the failures of experiments as necessary for learning is key; without that, most will be reluctant to be the first to "try something."

■ And last, but not least, have fun and laugh a lot!

Equally Informed People Seldom Disagree

I think this phrase resonated throughout our journey. The "Success is Assured" tools and methods are all about simplifying and visualizing information to help people come to consensus knowledge-based decisions. As employees become more specialized to address the complexities of their products, simple and effective collaborative methods are needed to keep them informed and aligned!

This concludes the case study of P&W; although it is still an ongoing story, there are some important lessons to build from. Those lessons will be covered in detail in the following chapters, but the most important are as follows:

■ Your organization can achieve immediate benefits from using visual tools to get its people equally informed.

■ The most critical time to do that is before key decisions are made prior to going into detailed design or the more expensive parts of your process (where rework is most costly).

■ However, your organization will not see sustained game-changing benefits just by giving your people the right tools and asking for "Success is Assured"; you cannot just add workload on top of the existing deliverables. The organization's leadership must stop asking for many of the old deliverables and time frames in order to make time for them to use the new tools, in order to identify and close the knowledge gaps that need to be closed to establish "Success is Assured" prior to making those key decisions.

P&W'S PURPOSE

> *We believe that powered flight has transformed and will continue to transform the world. So, we work with an explorer's heart and a perfectionist's grit to design, build, and service the world's most advanced and unrelenting aircraft engines, to turn flight's possibilities into realities for our customers.*

This purpose helped us be a trailblazer in adopting these principles. P&W didn't have a handbook telling them how to establish "Success is Assured." Much of their effort was trial and error; but it must be made easier. The experience described in this chapter provides a better understanding of the hurdles that must be overcome in order for teams to establish "Success is Assured." This has driven improvements to the teaching and coaching captured in this book. In continuing to work with many other companies on similar journeys, the authors have further developed the guidance we are sharing with you.

■ Chapter 2 will describe what we call "True North": a compass leadership can use in guiding the activity in their organizations. It then shows how to apply that compass in guiding the initiatives commonly found in organizations today (such as P&W's initiatives described in this chapter).
■ Chapter 3 then teaches the key enablers that the organization will need to identify knowledge gaps, rapidly close them, and ultimately establish that "Success is Assured" prior to making key decisions.
■ Chapter 4 will teach the Causal Mapping introduced in this chapter, which is an enabler of many of those enablers taught in Chapter 3.
■ But complex organizations like P&W have to collaborate across organizational boundaries, even across corporate boundaries. Chapter 5 will show how to align "True North" across organizational boundaries.
■ With that understanding, Chapter 6 will then give specific recommendations on many of the challenges and issues raised in this chapter and how to make the transition to these new practices in a complex successful organization.
■ Finally, Part II of this book will use a fictional story to give a concrete example of accelerating the required learning into the very earliest phases

of product development, with hopefully enough clarity that your organization will be able to do the same with your own problems in your own organizations.

The authors hope this look at P&W's journey will help make clear the reasoning for much of what is taught in the rest of this book and thus ease the adoption of these superior practices such that you can achieve the full benefits as rapidly and efficiently as possible.

Chapter 2

"True North"

It is quite common to see, as in the Pratt & Whitney case study, organizations putting in place many of the pieces (like K-Briefs, Causal Maps, Integrating Events, etc.), getting valuable benefits, but still having processes dominated by rework caused by bad early decisions. Why? In the end, it is not sufficient to talk about "Success is Assured" and adopt related practices; it has to change your decision-making. And very often, it is the same leaders asking for "Success is Assured" in one sentence who are unknowingly preventing it in another (e.g., asking for a standard deliverable by its standard due date, thereby forcing premature decisions). Why? In complex processes developing complex products, there are too many moving parts to understand how to change each and every one up front. Thus, it is critical that an organization establish clear guidance to their leadership on how they should be adjusting those moving parts when encountered. That clear guidance is the topic of this chapter.

Just as a ship's captain steering across a vast ocean will rely on a compass to point the way to true north in order to make better navigational decisions, a product development leader faced with a large variety of situations can more reliably keep his organization "on course" if he or she has a clear vision of "True North."

To provide such a clear vision of "True North" to product development organizations, we suggest the following more detailed statement of what we mean by establishing "Success is Assured" prior to decision-making:

> **We will adjust our processes, our behaviors, and our expectations to insist that we KNOW the impacts of the decisions we are making with enough clarity that we can make desirable trade-offs rather than committing our limited resources to decisions that will likely change.**

It is a statement that defines and insists on establishing that "Success is Assured" by consistently making knowledge-driven decisions. It states very clearly that making decisions without knowing all the ramifications and trade-offs is unacceptable behavior. Any process, tool, or method that doesn't support this thinking must be changed; any decision contrary to this will be challenged. In reality, we know that at times risks will have to be accepted and decisions will be made where success is *not* assured, but they should be actively minimized – not accepted as normal expected behaviors in order to stay on a published schedule. And further, in those cases, focused effort should be applied to close those knowledge gaps as quickly as possible as opposed to being satisfied to wait for the normal development process to reveal any limits that were violated.

As pointed out in Chapter 1, for an organization to change successfully, it is important for the leadership at all levels to have clarity on what and why they need to change from existing practices and take ownership of the changes. Without that clarity, they will invariably revert to the behaviors that they are most familiar with — as well they should. So, the litmus test for the preceding statement of "True North" is whether it will point to the desired behaviors in common situations. The remainder of this chapter will take the following common improvement initiatives in action across industry today, as illustrated in the P&W story, and see how they might change when guided by the "True North" statement. (*If they don't change, then we need a better "True North."*)

- Requirements Management
- Systems Engineering
- Phase Gate Processes
- Knowledge Management
- Problem-Solving
- Risk Management
- Lean (both Continuous Improvement and Waste Reduction)
- Application of the Toyota Product Development System (TPDS) principles

Requirements Management Guided by "True North"

Most Requirements Management initiatives are focused on three fronts:

1. Collecting the requirements from the customers
2. Allocating those out to subsystem teams and suppliers as subsystem specifications
3. Tracking, verifying, and validating the satisfaction of those requirements throughout that hierarchy

Those initiatives typically add rigor and structure to the process to make sure that all is captured in the front end such that it can be reliably verified and validated in the back end.

Although we traditionally think of requirements as inputs specified by the customer, in reality many of them are part of the learning we must do during the product development process. Most customer requirements are overconstrained (i.e., they cannot all be achieved simultaneously). In fact, trade-offs will need to be made directly by the customer, on behalf of the customer, or ideally in collaboration with the customer. Thus, to what degree you will choose to satisfy each of the "requirements" become key decisions — often the most critical decisions to be made in the development process!

In contrast, we typically give those critical requirements decisions relatively little thought. We simply accept them as already made — decided by the customer — inputs to the process, rather than outputs of the process. For that reason, we recommend calling them *customer interests* — or better, *customer interest decisions* — rather than *requirements*, so that people start embracing them as the decisions that they are.

So, applying "True North" to Requirements Management leads you to focus on identifying the key customer interest decisions and then identifying the knowledge gaps that need to be closed in order to make those decisions with "Success is Assured." In contrast, Requirements Management efforts typically focus on the early identification of all levels of customer requirements — a shopping list of sorts to ensure nothing is forgotten. More important is to apply an initial focus on the critical customer interests that it may not be feasible to meet and then allow the more routine requirements to flow out later within the learning process.

To do that, you need to make visible the capability limits that force the trade-offs between the customer interest decisions. Those are the first knowledge gaps that need to be closed. Once the trade-offs are visible, only then can you start to work on closing the second set of knowledge gaps: where do the customers want the product to be on those trade-off curves? That learning is generally not possible without visibility to the trade-off curves.

Traditional Requirements Management often tries to establish relative priorities between the requirements, in case they cannot all be satisfied. But it is very difficult for customers to give such prioritizations accurately because the levels matter in that prioritization. For example, reducing the size of cell phones was critically important when they were carried around in the trunk of a car or a backpack. But now that they easily fit in your pocket, making them smaller is of zero importance. In fact, they started trending bigger again to maximize screen size.

When making trade-offs, it is not about relative priority. It is about the shapes of the trade-off curves. How much X do I have to give up to get more Y? And it is about specific levels to satisfy specific use cases. It is those levels and those trade-off curves that need to be made visible in order to truly learn what the customers really want. The illustration of that in Figure 2.1 will be explained in detail in Chapter 3.

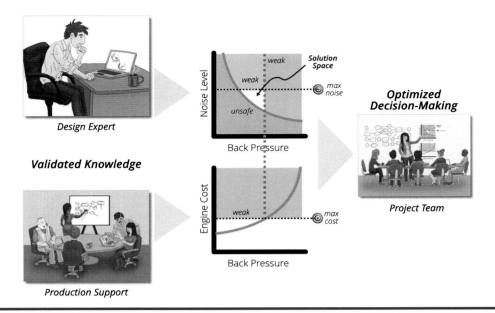

Figure 2.1 Optimized decision-making by combining Set-Based knowledge.

Whether designing a cell phone, a laptop computer, a race car, or a jet engine, the team will be working hard to reduce weight and reduce cost. But often, the options to reduce weight involve more cost; or options to reduce cost increase weight. So, what is better? You can't just go ask your customer, "What's more important, less cost or less weight?" They will invariably answer, "Both!" Rather, you need to show them the actual trade-off: "I can reduce cost from C to D by increasing weight from W to X. Would you prefer a product that costs C and weighs W or that costs D and weighs X?" Or better yet, show them the full trade-off chart of cost versus weight and let them choose where they want to be in that design space.

So, with a focus on "True North," Requirements Management initiatives should be putting in place early learning that first makes visible the causal connections between the different customer interest decisions, identifying the knowledge gaps regarding the sensitivities between them that need to be closed in order to make the trade-off curves and levels visible. And that learning must be followed with additional learning focused on where the customers want to be on those trade-off curves. To make the most desirable decisions, such that those customer interest decisions will not need to be remade, quite a bit of learning and analysis needs to be done. And that's just to make the decisions that we normally have at the very beginning of the process (as if they were inputs). In fact, they are probably the last decisions made before establishing "Success is Assured" for the larger design.

If your Requirements Management initiatives are not putting in place such rapid learning to help you converge to the optimal set of customer interest decisions prior to the start of detailed design (the latest point that we want to have established "Success is Assured"), then your Requirements Management initiatives are not being guided by "True North."

Systems Engineering Guided by "True North"

Many companies use Systems Engineers to manage the Requirements Management initiatives covered in the prior section. So, in this section, we are looking at all the other areas of Systems Engineering, ignoring Requirements Management.

For those of you that do not have a Systems Engineering group in your company, here is the International Council on Systems Engineering (INCOSE) definition of Systems Engineering:

> *Systems Engineering (SE) is an interdisciplinary approach and means to enable the realization of successful systems. It focuses on defining customer needs and required functionality early in the development cycle, documenting requirements, then proceeding with design synthesis and system validation while considering the complete problem: operations, cost and schedule, performance, training and support, test, manufacturing, and disposal. SE considers both the business and the technical needs of all customers with the goal of providing a quality product that meets the user needs.*[1]

A complex product consists of a group of subsystems working together to form a complete system. The design synthesis function of a Systems Engineering group involves the integration of all the subsystems into a successful product.

Typically the Systems Engineering group will parcel out the customer requirements to the subsystem teams. The subsystem teams then initiate their design processes. When a subsystem determines they can't meet the requirements, they request relief from the Systems Engineering group, who then look across all the subsystems to see where they can "rob Peter to pay Paul." Discussions become heated when one subsystem has to accept a more challenging requirement so the other subsystem has a chance of meeting its requirements. This happens over and over again throughout the development process. Often, the customer is contacted to see if they will accept a requirement miss — not a pretty discussion. Imagine if the Systems Engineering group finds out that a requirement isn't achievable at a design review. All the other subsystems are well into their design process, making requirement adjustments costly and time-consuming. The net result is an extremely inefficient and unstable process that is hard to control and delivers unpredictable results.

When applying "True North" to Systems Engineering, the Systems Engineering group should be identifying the key interface decisions between subsystems up front and then making visible the trade-offs between those interface decisions as dictated by the constraints being actively uncovered by the subsystem teams.

---

[1] INCOSE, *Systems Engineering Handbook: A Guide for System Life Cycle Processes and Activities*, 4th Edition, edited by David D. Walden, ESEP, Garry J. Roedler, ESEP, Kevin J. Forsberg, ESEP, R. Douglas Hamelin, Thomas M. Shortell, CSEP, Hoboken, NJ: John Wiley, 2015, p. 265.

With those connections visible, they can proactively lead the subsystem teams to identify any knowledge gaps that need to be closed in order to establish that "Success is Assured" for those key interface decisions (much like with the customer interest decisions).

While those knowledge gaps are being closed, the Systems Engineering group should be resisting any premature decision-making that would compromise their flexibility. However, as those knowledge gaps are closed in one subsystem, they may learn that some portion of the range is undesirable and/or that some portion of the range is infeasible. The infeasible portions can be eliminated as weak immediately, and that convergence of the interface decision to a smaller range should be communicated to the other subsystem teams immediately, so that they can narrow their focus. The less desirable ranges may need to be retained in case other subsystem teams determine that the more desirable ranges are infeasible. It is the Systems Engineering team that should be monitoring those interface decisions, understanding what knowledge gaps remain, and therefore knowing which weak parts of the design space should be eliminated sooner rather than later.

As those interface decisions converge, they will be driving convergence in the customer interest decisions. That convergence may trigger the need to show the customer the newly exposed trade-offs, as discussed in the prior section on Requirements Management. From the customer, the Systems Engineering team will learn what parts of the design space are weaker and should be eliminated, which is then communicated back to the subsystem teams.

In Chapter 3, we will discuss a number of key enablers that will help the systems engineers manage this ongoing convergence process in cooperation with the subsystem teams — in particular, the Trade-Off Chart and the associated Decision Map, which allow the subsystem teams to make visible the limits on the interface decisions imposed by their subsystem in a way that can be understood by everyone, without needing deep expertise on their subsystem. Further, those enablers allow those limits to be projected out to the customer interest decisions such that those trade-offs can be made visible to the customers.

If your Systems Engineering initiatives are not putting in place such visibility and not enabling a convergent, collaborative, and concurrent decision-making process, then your Systems Engineering initiatives are not really being guided by "True North."

Phase Gate Processes Guided by "True North"

In Pratt & Whitney's industry, the phases, gates, and gate criteria are largely dictated by the Department of Defense customer on the military side, and similar phases and gates tend to be used as best practices on the commercial side. The different prime contractors and suppliers may add additional gates and additional criteria to help improve their own performance, but the bulk of the process is dictated by the customer.

Those phases, gates, and gate criteria do not get in the way of "True North," but they also do not support or pull it. The danger is that satisfying the gate criteria may distract focus from what is more important. In particular, it puts the focus on completing tasks or deliverables rather than on closing knowledge gaps. The focus on completing tasks tends to put pressure on making decisions; the lack of focus on closing knowledge gaps can then result in decisions being made prior to the knowledge being available to make those decisions such that "Success is Assured."

In solving this, there will be a tendency to just add gate criteria regarding knowledge gaps and decision convergence without removing any of the existing criteria. And that may be the right thing, depending on your existing criteria. But leadership needs to be sensitive to the fact that

1. You can't just add more work to fully loaded staff without consequences; you need to look for things to offload.
2. You can't put focus on something without taking focus off something else.

Where this is most critical is in the earliest gates. Those often have deliverables that are forcing premature decisions. In those cases, adding on knowledge learning and evaluation activities will not be enough. Those deliverables forcing premature decisions really need to be pushed out to later gates (possibly newly established gates), thereby making more time for the learning to close knowledge gaps and converge to "Success is Assured" decisions.

Possibly an easy place to offload is gate criteria associated with Risk Management. Most of such risk is due to knowledge gaps. Rather than manage that risk through tracking and monitoring at various gates, it is better to eliminate it completely by closing the knowledge gaps. You may still need some Risk Management where you can't close knowledge gaps, but if you can eliminate three-quarters of the risk, that alone will reduce the Risk Management workload. But better, as total risk drops, the level of interaction between different sources of risk drops, and thus the impact drops exponentially. That coupled with the flexibility that the added reusable set-based knowledge brings you when reacting to surprises may allow you to relax some of the rigor and demands of your Risk Management process. (We'll elaborate more on Risk Management later in this chapter.)

If your Phase Gate initiatives are not shifting the focus to closing the knowledge gaps to establish "Success is Assured" for the key decisions, then clearly they are not really being guided by "True North." When evaluating that, don't be fooled by gate criteria that simply indicate that success should be assured without identifying the key decisions or discussing the required learning. Too often we expect that executing the traditional tasks to generate the traditional deliverables will assure success. As long as those can be based on faulty knowledge, opinions, guesses, or other wishful thinking, success is *not* assured! Identifying and purging such assumptions needs to be the focus of the phase gate initiatives guided by "True North."

Knowledge Management Guided by "True North"

Most companies have several Knowledge Management systems in place, but many can be best described as "black holes" (lots of knowledge is captured, but very little knowledge is ever taken back out except by the person who put it in), and almost all have had disappointing levels of knowledge reuse.

Many of the Knowledge Management initiatives are working to remedy that by improving the search tools, categorization, ontology, and so on. Many more of the Knowledge Management initiatives, not finding success there, have started looking at the larger system of Knowledge Management, focusing on establishing Communities of Practice, Expert Directories, Knowledge Brokers, Knowledge Supermarkets, and so on.

However, none of these efforts are getting to the root cause of the lack of knowledge reuse: the lack of knowledge *use*. In the end, if the knowledge you are managing does not affect the decisions you are making, then it is having no positive effect. And if the way you make decisions is not based on knowledge, then there really isn't an effective path for knowledge reuse. First, you need to change how you use knowledge to make decisions, and then your Knowledge Management needs to be redesigned to support that.

So, assuming "True North" leads us to a decision-making system based on knowledge, what then should our Knowledge Management system look like? The ideal Knowledge Management system should require zero extra effort. Nobody wants to manage knowledge. Nobody wants to capture knowledge, other than what is necessary to get the job done. Nobody even wants to search for knowledge; they want to *have* that knowledge in order to save themselves work in getting that knowledge, but they don't want to have to search for it.

If a team has its knowledge baked into its best-practice decision-making tools, but then they discover new knowledge, then the proper course of action should be to first incorporate the new knowledge into their best-practice tools such that they can then make (or re-make) the decisions for this project using those revised decision-making tools. In that way, they ensure all past knowledge is still being respected in that new decision. And as a side-benefit, that knowledge is captured forevermore in their best-practice decision-making tools.

In other words, the easiest way to get my current project done captures the knowledge right where I want it to be to reuse it: in the decision-making tool. Thus, no extra work is required for knowledge capture beyond what I'd have done for this project anyway. And Knowledge Management is then really just the management and continuous improvement of our team's best-practice tools for making the decisions that team is responsible for making.

If your Knowledge Management initiatives are not starting with and focused on how knowledge is *used* to make decisions and finding ways to support that, then they will not achieve knowledge *reuse*, and they are not really being guided by "True North." And if those initiatives are insisting that all knowledge be housed in some standardized system that is not part of the decision-making process, then that initiative might actually be getting in the way of "True North" or at least a major distraction from it.

Problem-Solving Initiatives Guided by "True North"

Given 65%–75% of your engineering capacity is probably consumed with firefighting, it makes a lot of sense to invest in initiatives focused on improving the efficiency of your organization's problem-solving. Often problem-solving initiatives come as part of other initiatives that have a strong problem-solving element at their core. Others are focused on specific Problem-Solving or Root Cause Analysis methodologies. Here are a few of the common ones:

- Eight Disciplines (8D) Problem-Solving
- Root Cause and Corrective Action (RCCA)
- Kepner–Tregoe Problem-Solving
- A3 Problem-Solving
- Plan-Do-Check-Act (PDCA)
- Observe, Orient, Decide, Act (OODA)
- Design, Measure, Analyze, Improve, and Control (DMAIC; under Six Sigma or Design for Six Sigma)
- Theory of Inventive Problem-Solving (TRIZ; Russian acronym)

Many companies have created their own Problem-Solving methodologies based on one or more of these or others, tailored to their own needs.

However, even where structured problem-solving or root cause analysis has been put in place, there is still a high percentage of problem fixes that end up causing other problems or effectively changing the trade-offs to something undesirable. As long as fixing problems creates other problems, the backlog of problems will never end and the firefighting will continue to consume the majority of your engineering capacity.

So, with "True North" as a guide, the focus should be not only on the decisions that need to be changed to fix the problem but also on the knowledge that needs to exist to prove that "Success is Assured" — meaning we won't cause other problems with our solution. In other words, "True North" is not just a guide for the critical decisions made early in projects, it is also a guide for the decisions being remade late in product development to fix the problems. These may be much smaller decisions but still important enough that they are consuming the majority of your engineering capacity.

For many organizations burdened by large problem backlogs, the application of "True North" to the problem-solving efforts comes first, and that allows them to rapidly work down their problem backlogs, freeing up capacity that they can then apply to new projects, where they can then start applying "True North" for even bigger benefits.

If your Problem-Solving initiatives are focused primarily on problem descriptions and root cause analysis, but not focusing on the impacts that changing those root causes may have on the larger trade-off decisions in your product designs and the other problems that might be caused, then those problem-solving initiatives are not really being guided by "True North."

Risk Management Guided by "True North"

Risk Management isn't typically an initiative being pushed to make improvements as much as it is just an ever-growing presence in Product Development management, as the consequences of that "risk" continue to grow with the complexity.

So, what does "True North" say about risk?

Stop thinking of it as a risk; that just leads people to think about it statistically, glossing over the details of what it really is: knowledge gaps. Instead, identify the knowledge gaps that are the root cause of the risk, and get focused on closing those knowledge gaps. Each time you do, you not only "burn down" that risk, you do so permanently. But more importantly, you can now see the business trade-off:

- You can estimate the cost of closing each knowledge gap.
- You can estimate the delay (if any) of closing each knowledge gap.
- You can estimate the potential impact of each knowledge gap.
- You can see the potential interactions of each with the others.

That gives the decision-makers much more information on whether or not to close the knowledge gaps rather than simply using a "gut feel" statistical evaluation of risk levels.

More importantly, it gets your engineers thinking and innovating on cheaper or faster ways to close those knowledge gaps, at least sufficiently to reduce the potential impact or the potential interactions. And those potential interactions guide which knowledge gaps are most critical to close (because it's usually the interactions that bite you; or said another way, where knowledge gaps interact, the real risk tends to be squared, not additive, so you burn down far more risk whenever you can break those interactions).

Some will push back that some things are not knowable. For example, you cannot know the price of oil or the weather conditions that you might encounter 10 years from now. But establishing "Success is Assured" does not require you to know that; rather, you just need to know the worst case (the extremes) and then establish that your solution can handle those extremes — that your solution can handle the full set of possible situations. That tends to be a much simpler analysis. Bounding the extremes is typically a knowledge gap that can be closed, even when the particular values will never be known in advance.

Where you can't close knowledge gaps, you may need to fall back on traditional Risk Management. But that should always be a last resort. Focus on the root cause: the knowledge gaps!

If your Risk Management is perceived as the primary method for dealing with the potential consequences of your decision-making-with-inadequate-knowledge, rather than focusing on identifying and closing the knowledge gaps to avoid the risk entirely, then your Risk Management is not being guided by "True North." The best Risk Management is having no risks to manage ("Success is Assured").

Lean Guided by "True North"

Much of the "Success is Assured" decision-making methodology was derived from our eight years of working with Dr. Allen Ward based on his research on the TPDS; so, it should be no surprise if Lean and "Success is Assured" are highly complementary. However, many of the Lean initiatives in product development are based heavily on what worked in Lean Manufacturing. When that happens, the results tend to be disappointing for the reasons outlined in the P&W story.

Although it is an oversimplification, the three common themes of Lean initiatives are

1. Waste Reduction
2. Cadence, Pull, and Flow (minimizing batch size; minimizing work-in-progress)
3. Continuous Improvement

Let's apply our "True North" to each.

The biggest waste in most product development organizations is rework. Many companies have measured that 65%–75% of their engineering capacity is consumed doing rework. Eliminating rework alone represents a three- to four-fold productivity improvement. But Lean principles would argue that waste is even more costly than that. Given the enormous size of that waste, most Lean initiatives' Waste Reduction efforts should be focused on reducing rework, unless another comparably sized waste is identified. (Reducing clutter on engineers' desks or optimizing the steps of a paperwork process are probably not worth the distraction as long as the four-fold improvements from reducing rework have not yet been realized.)

Another huge waste in most product development organizations is the re-creation of knowledge (failure to reuse existing knowledge). Failure to reuse knowledge also impacts the *Continuous Improvement* objective of Lean. Without effective knowledge reuse, there are two negative effects on continuous improvement: you don't have your knowledge in a form that it can be challenged and improved, and even if you did improve your knowledge, it would rarely be reused, so there would be little benefit.

When we apply "True North" to Continuous Improvement, the focus of improvement should be on our ability to establish that "Success is Assured" prior to making decisions. To better do that, *beyond Continuous Improvement of our reusable knowledge*, we would want to continuously improve

■ Our ability to identify the gaps in the knowledge needed to make decisions
■ Our ability to learn deeply and to quickly close those knowledge gaps
■ Our ability to make tough trade-offs during decision-making

For establishing proper Cadence, Pull, and Flow to support "True North," we need to focus on the rapid learning activities that feed the decision-making. Since there will always be more to learn than we have time for, we need to use the required timing of the decisions to pull the learning for the specific knowledge gaps that need to be closed to make those decisions. By using the key decision points to pull the learning, we keep our organizations focused on the most important knowledge gaps.

With that pull established, you want to leverage the early learning to drive the key decisions that "eliminate the weak." As you converge the decisions, some of the knowledge gaps will no longer need to be closed (minimizing the work required to do all the required learning); thus, you need to put into place a rapid cadence of planning what should be in each next focused learning effort. Tools and mechanisms to feed such rapid convergence will thus be important.

If those are not the sorts of things your Lean initiatives are doing in product development, then they are not being guided by "True North."

The Toyota Product Development System *is* Guided by "True North"

Some Lean initiatives in product development have rightly focused more on the TPDS than on the Lean Manufacturing principles seen in the Toyota Production System (TPS). For the last two decades, the TPDS has been the model of product development excellence and rightfully so; their system routinely creates highly reliable, profitable cars that people want and in less time than their competitors. They also consistently meet production schedules with minimal loopbacks.

Many books have been written describing how they do it. Most are focused on observed practices and promote copying their practices to achieve Toyota's level of excellence. The assumption is they have been developing the practices for decades, so just copy them. The reality is the Toyota practices are closely aligned to their culture, their product, and their industry and its customers. It has not proven to be a system easily replicated in other engineering environments.

Rather than trying to copy their practices, it may be more instructive to look at how those practices enable "True North" and then use "True North" to similarly guide your own practices.

From the perspective of the initiatives guided by "True North" that we just discussed, the TPDS meets all the "True North" characteristics:

■ Toyota focuses on closing the knowledge gaps regarding both the real objectives of the customer and the limits of their capabilities so that they can make the right trade-offs before committing to a project.

■ Systems engineering at Toyota allows for ongoing convergence to an optimized set of targets based on understanding and resolving limitations and trade-offs.

■ Their scheduling and control mechanisms are based on planned decision points and validated visible knowledge that supports the decisions.

■ By capturing their knowledge in the form of the checksheets they use to make their decisions, reuse and continuous improvement of the knowledge is almost unavoidable.

■ Their A3 problem-solving process supports knowledge-based decision-making at every level.

■ Risk management is minimized through their relentless focus on identifying and resolving knowledge gaps.

Allen Ward once made an interesting observation that, at Toyota, product development was not primarily focused on designing cars but rather focused on creating knowledge about cars; a cadence of great cars would result from that knowledge.

Their product development system is actually quite simple, based on the continuous flow of knowledge to make decisions. First, let's look at the two key Toyota leadership roles and their responsibilities:

1. Chief Engineers
   a. *Roles*: They are the advocates of the customer and they are the product designers. They start the process with high-level targets and take time to understand the customer interests driving those targets. Then they identify the key decisions that need to be made and set a cadence of reviews to understand the functional capability limits, which are then used to converge the targets into design specifications. They pull the learning process to make the design decisions that need to be made.
   b. *Responsibilities*: They set the schedules for key decisions and make the narrowing design decisions based on the knowledge provided by the Functional Managers.
2. Functional Managers
   a. *Roles:* They manage and mentor the engineers through the learning process in order to provide the knowledge to the chief engineer to make the key decisions.

b. *Responsibilities:* They provide knowledge of adequate quality to make the decisions that need to be made. They are responsible for establishing, improving, and teaching that knowledge in an ongoing way.

Given that split of responsibilities, some level of conflict is expected and, in fact, desired, as it tends to drive more critical thinking and innovation. As one Chief Engineer commented, "Great conflict creates great cars." But that is only true if that conflict is resolved based on validated knowledge, not on opinion or wishful thinking.

Figure 2.2 is a simple visual model of this interaction.

This is the heart of the TPDS, and it is a system optimized for achieving "True North." The Chief Engineer has the explicit role of making visible the targets for the customer interests and making the trade-off decisions on behalf of the customer for whom they are advocating. The Functional Managers have the explicit role of providing the knowledge necessary to make the trade-offs visible so that the Chief Engineer can make those trade-off decisions. The Chief Engineer lays out the timing of those decisions; the Functional Managers allocate the engineering resources to build the knowledge by that time. The inevitable conflicts from those overlaps create an environment for innovation focused on the best way to, in essence, achieve "True North."

In trying to adopt TPDS practices, you need not copy Toyota's organization or specific behaviors. Rather, focus on establishing a system that ensures an ongoing focus on "True North" — on establishing that "Success is Assured" prior to making decisions. If instead you are copying specific practices, but those practices

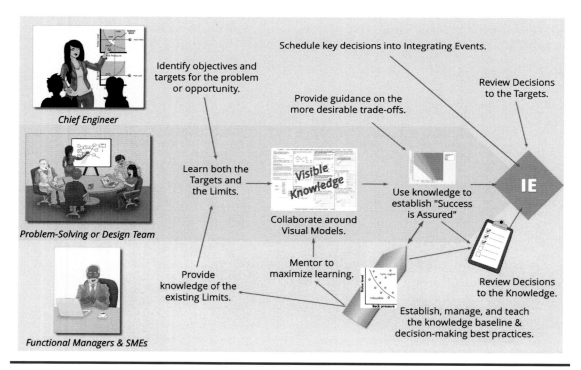

Figure 2.2   Key roles and interactions leading to an Integrating Event.

are not establishing that "Success is Assured," then you may not see the benefits that you are hoping to see.

Product Development Guided by "True North"

Hopefully, this look at common initiatives in product development organizations and what it means for them to be guided by "True North" will help you reevaluate not only your initiatives but all your product development processes more effectively. If we aren't identifying the key decisions and the knowledge required to make those decisions, then we are fooling ourselves that we are on a path to establish "Success is Assured." We must innovate ways to close the knowledge gaps prior to making the key decisions or innovate ways to delay those decisions until the knowledge can be acquired.

Finding ways to establish that "Success is Assured" prior to making decisions (or delaying decisions until "Success is Assured") is itself an acquired skill set.

- Product development leadership needs to develop the mastery of demanding that "Success is Assured" and reviewing the supporting knowledge.
- Engineering needs to develop the engineering capabilities to deliver the knowledge that establishes that "Success is Assured."

This chapter has been about the former; the next three chapters are about the latter, and then Chapter 6 will return to a focus on the former, taking a closer look at the challenges that leadership will need to overcome.

Chapter 3

The Enablers

If the ship's captain accurately points to "True North," but nobody on the ship knows how to configure the sails to get the ship moving in that direction, then knowing "True North" will not have much value. Given the potential for four-fold productivity gains and development cycle times cut in half, all with higher quality and greater innovation, it is those who master the skills to achieve "True North" that will dominate their industries in the future. But it is not good enough to simply lead the way to achieving "True North," you must equip your organizations to achieve it.

This chapter will look at the enablers for establishing that "Success is Assured" prior to making key decisions.

Are these the only possible enablers? Absolutely not! Working with customers across a variety of industries, we continue to develop new enablers to this day.

Are all of these enablers required? Not necessarily. However, they each build on and depend on the others. So, you may be able to leave some out, but not without filling in that void with some other enabler. The one enabler that is absolutely essential is Set-Based — in complex decision-making, establishing "Success is Assured" will be impossible without making the move to Set-Based.

Do these enablers have a sequence? (Can I start with just one or two?) There is no particular sequence for adopting them; they each build on and depend on each other. But they each have their own isolated benefits. P&W's use of Causal Mapping to solve a weight issue is a good example of this. So, you can start with those that will have the most near-term benefits to your organization by themselves and then add on the rest to achieve the larger benefits of establishing "Success is Assured."

Challenges in Achieving "True North"

In order to achieve "True North,"

We will adjust our processes, our behaviors, and our expectations to insist that we KNOW the impacts of the decisions we are making with

enough clarity that we can make desirable trade-offs rather than committing our limited resources to decisions that will likely change.

What difficulties and challenges must engineers and decision-makers overcome? In modern complex product development, here are some of the bigger challenges:

- As complexity rises, the need for knowledge from multiple people, and thus the need for effective collaboration, rises.
- Further, as complexity continues to rise, the need for specialists in different areas arises, which then requires collaboration across areas of expertise — meaning the collaborators don't necessarily fully understand each other.
- As products mature and improve, they push the limits more, increasing the number of trade-off decisions that must be made, increasing the dimensionality of the design space to be analyzed.
- As competition rises, product specialization rises, and the requirements become more specific and more demanding, leading to more trade-offs and more challenging trade-offs.
- As governments add regulations, as safety expectations rise, and as security expectations rise, the number of requirements rise, and thus the number and complexity of trade-offs that must be made rise.
- As the number of trade-offs rise and become more challenging to meet, the level of interconnection will tend to rise, such that design decisions tend to impact more and more trade-offs in the multi-dimensional trade-off space.

How Big Are These Challenges?

Bad commercial results tend not to be widely published; so, let us look at reports to the U.S. Congress on their development acquisitions.

The U.S. Government Accountability Office (GAO) reported that the 96 major U.S. Department of Defense (DOD) programs in progress in 2009 had a cumulative cost overrun of $296 billion (over $3 billion per program) and an average schedule overrun of 22 months. They indicated agreement with the DOD on the sources of the problem, including inadequate knowledge, excessive requirements, an imbalance between wants and needs, and permitting the programs to move forward with such. They called for programs to "follow a knowledge-based approach," to "begin with strong systems engineering analysis that balances a weapon system's requirements with available resources," and to use "a knowledge-based approach to product development that demonstrates high levels of knowledge before significant commitments are made."[1] Sound familiar?

Despite concerted efforts to contain such cost and schedule growth since then, in 2016 (seven years later), the GAO reported that there were 79 major

---

[1] U.S. GAO, Assessments of major weapon programs, GAO-09-326SP, March 2009. Online: http://www.gao.gov/new.items/d09326sp.pdf. Accessed April 12, 2017.

programs in progress, with a cumulative cost overrun totaling $469 billion (over $5.9 billion per program) and an average schedule overrun of 29.5 months. The GAO found that while a few programs are using knowledge-based approaches, the application has been uneven, and they call for improvements as "knowledge supplants risk over time."[2] (That last quote also appears in the 2009 report… but, evidently, was not yet realized seven years later.)

In yet another report, they make the statement, "Despite decades of reform efforts, these outcomes and their underlying causes have proven resistant to change and, in fact, both DOD weapon system acquisition and DOD contract management have been on our high-risk list for nearly 20 years."[3]

We suggest that identifying the need for better knowledge prior to making commitments in order to supplant risk is not enough; after all, most (if not all) the people on those 79 programs would prefer to make knowledge-based decisions and would prefer to use knowledge to supplant risk. The problem is that the knowledge is not available when they are required to make the decisions in the front ends of those programs. The GAO should instead be calling for the DOD and its suppliers to learn a new way to operate in the fuzzy front end of those programs that will accelerate the required learning ahead of the required decision-making or (if that is not possible) be willing to delay the required decision-making until that knowledge can be learned. Accelerating that learning is where the enablers described in this chapter come in; existing tools and best practices are not up to the task.

Accelerating Learning Is *Not* Easy!

We are certainly not the first to suggest that we need to find ways to accelerate our learning in the earliest phases of product development to avoid very expensive rework. Nor are we the first to suggest ways to accelerate that learning. The most recent trend in those suggestions involves adopting *Agile* practices that have been successful in software development, such as stand-up meetings and Scrum-like *sprints*, where the work is learning activities rather than software development. And while those are generally beneficial, and we advocate some of that as part of our recommendations, those practices will not provide the level of acceleration being called for here.

For complex product development, where the learning normally occurs based on analyses of the detailed designs, typically CAD models, no amount of acceleration of that existing learning process is going to make that knowledge available prior to the *start* of that detailed design, let alone even earlier in the process, when key decisions need to be made based on that knowledge.

[2] U.S. GAO, Assessments of major weapon programs, GAO-16-329SP, March 2016. Online: http://www.gao.gov/assets/680/676281.pdf. Accessed April 12, 2017.

[3] U.S. GAO, Managing risk to achieve better outcomes, GAO-10-374T, January 2010. Online: http://www.gao.gov/assets/130/123946.pdf. Accessed April 12, 2017.

To accelerate that learning ahead of those decisions (to achieve "True North"), the learning must be accomplished in a very different way. But it still needs to be based on collaboration across most of the same expertise boundaries. It still needs to uncover all the key influencers across domains of expertise. Any other decision that may have a significant influence has to be identified and the knowledge regarding the level of influence acquired. If the known process to do that takes months or years, but needs to be accelerated into weeks or even days, a fundamental change in approach is needed.

In evaluating an approach to accelerated learning, consider how it is reducing workload! In the end, our engineers are working hard to get things done fast; simply telling them to go faster is not going to help. Agile and similar methods can improve efficiency, but that is a percentage gain, not a massive multiplier. Minimizing their workload has to be an integral part of the approach.

Set-Based Can Be the Accelerator

Set-Based is generally accepted as an enabler of delaying decisions until you can establish the knowledge needed to make the decisions. However, Set-Based is typically not associated with accelerating the learning (the preferable approach to achieving True North). When characterizing Set-Based practices, many people emphasize the carrying of additional options. If that was it, then Set-Based would be adding workload, not reducing it, and thus would not be an effective accelerator. And one of the most commonly cited downsides of Set-Based is that you pay more up front, but that is justified because it pays off later.

In contrast, we argue that Set-Based practices, leveraged properly, are a powerful mechanism for minimizing workload and increasing concurrency, and thereby accelerating learning.

- Set-Based allows you to just analyze the worst cases for the design space and prove that the worst case is good enough to establish "Success is Assured," reducing the precision that you need to compute early in the process.
- Set-Based allows you to compute in terms of ranges of values such that you can leverage the fuzzy knowledge that you already have early in the process.
- Set-Based makes the sensitivities and limits of the design space visible in such a way that it is easily shared across areas of expertise. This allows you to analyze the broader impacts of the decisions being converged.
- On the flip side of that benefit, Set-Based allows teams working in different areas to work concurrently, as they steadily converge decisions in parallel.
- Set-Based allows you to determine the rough slopes (sensitivities) without needing to know the precise levels, giving you early visibility to what parts of the design space will be better or worse.

- With visibility to those sensitivities and the causal relationships among the decisions, Set-Based allows you to focus on the key decisions that have the greatest impact, thereby simplifying the design space that must be addressed early on.
- As a corollary to that last point, those sensitivities also make clear the decisions that have minimal impact, making clear the decisions that your team should not waste time dwelling on. (Without that visibility, such decisions that lack a strong driver can end up consuming the most debate time.)
- Set-Based allows you to converge decisions steadily toward better parts of the design space without needing to precisely analyze particular points, thereby avoiding the multiple iterations often required by those point-based analyses.
- Set-Based guides you to use what you know to quickly eliminate the weaker parts of the design space (such "eliminate the weak" decisions tend to be far easier to make than "pick the best" decisions).
- Further, by quickly eliminating the weak, you quickly reduce the quantity of design space that must still be learned, saving more time.
- Given Set-Based allows you to converge decisions over time, it reduces the early burden of trying to select specific point values for those decisions; you only need to narrow the decisions (eliminate the weak).

In the U.S. Navy's first attempt at working a Set-Based program, leveraging not much more than about half of the benefits of Set-Based,[4,5] the Ship-to-Shore Connector program finished preliminary design on time (despite having a more aggressive schedule than normal, which was the motivator for trying Set-Based), finished only 10% over the original budget, and used none of their design margin (i.e., had no need for loopbacks)[6] — all dramatically better than any other Navy program. Perhaps as Set-Based continues to be more widely adopted by Navy programs, the GAO will start to track the numbers presented earlier separately for programs that operate Set-Based or focus on establishing "Success is Assured."

Altogether, Set-Based allows you to focus on

- The key decisions
- The key relationships that need to be understood to evaluate those decisions' impact on your targets
- The subset of the design space where those relationships need to be understood
- The specific gaps in your knowledge that are preventing you from narrowing those key decisions (eliminating the weak parts of the design space)

---

[4] T. A. McKenney, M. E. Buckley, and D. J. Singer, The practical case for Set-Based design in naval architecture, paper presented at the International Marine Design Conference, Glasgow, UK, June 11–14, 2012.

[5] T. A. McKenney, An early stage Set-Based design reduction decision support framework utilizing design space mapping and a graph theoretic Markov decision process formulation, PhD diss., University of Michigan, 2013.

[6] Norbert Doerry, Naval Set-Based design, presentation, March 1, 2017. http://www.doerry.org/norbert/papers/20170223SBDdoerry-60minutes.pdf. Accessed August 14, 2017.

And in that much narrower focus, it allows you to use the lowest precision analyses that are adequate to establish that "Success is Assured" for the worst case in the remaining design space.

In other words, Set-Based greatly reduces the workload in the front-end phases through what we call *Decision-Focused Learning*.

Decision-Focused Learning

Figure 3.1 depicts the fundamental ingredients in product development decision-making. First, we have some *Targets* that we are trying to achieve, including both our business objectives and the customer interests that we will need to satisfy to achieve those business objectives. And then we typically have a great many *Ideas* on how we might achieve those Targets, including ideas for small incremental improvements, as well as ideas for fundamentally different approaches to satisfying the customer. But in the end, we must obey the physics of the real world. There are fundamental limits to our capability to deliver on those Ideas and to achieve those Targets. Given that, it is almost universally true that we will not be able to achieve all the Targets, and thus *Trade-Offs* must be made.

Ultimately, development teams must decide to what degree they are going to satisfy each of the Targets (the customers' requirements, where the business itself is one of those customers). To do that, they must decide what Trade-Offs to make on behalf of those customers. To do that, they must decide which Ideas to use based on each of their *Capability Limits*, given the underlying decisions that must be made in the design of each Idea.

However, early in product development, there are gaps in the knowledge of each of those. More importantly, there are gaps in the understanding of the interconnections. Ultimately, it is the Capability Limits that force the Trade-Offs, and it is those Trade-Offs that will drive the decision-making on which Ideas will best satisfy the Targets, and to what degree those targets can be satisfied in concert.

Figure 3.1 Fundamental ingredients in product development decision-making.

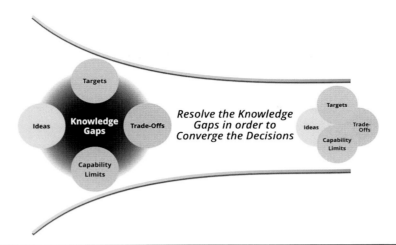

Figure 3.2   Knowledge gaps must be closed prior to decision-making.

The key point is that knowledge gaps from any of the five perspectives (Targets, Ideas, Capability Limits, Trade-Offs, or interconnections) must not be left undiscovered (Figure 3.2); *if you don't discover them early, even for problem-solving, they will eventually make themselves known at a much greater cost for resolution.*

After some practice at identifying the knowledge gaps that prevent making decisions such that "Success is Assured," people will often ask us, "Okay, I see the value of making visible all our knowledge and our knowledge gaps, but what about the gaps we don't know about — the *unknown unknowns?*"

The right visual models can actually help you see the existence of unknown unknowns. They help you see that a trade-off must exist at this spot in the model, pulling the right set of questions to ask, such that those unknown unknowns are uncovered and added into the model. We will discuss an enabler of that, the Causal Map, in the next chapter. Figure 3.3 adds the Causal Map as the tool for connecting the capability limits to the targets in order to expose the Trade-Offs, and do that in a way that even the unknown unknowns are exposed.

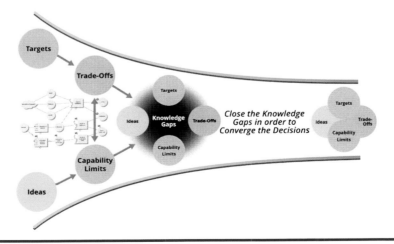

Figure 3.3   Causal Maps help make knowledge gaps visible.

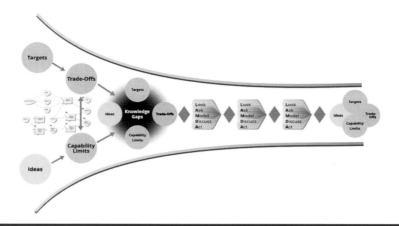

Figure 3.4 Decision-Focused Learning process.

With the knowledge gaps quickly identified, the team can focus on rapidly closing those gaps, and as they do so, rapidly narrowing the focus on what should be closed next. That results in the Decision-Focused Learning process depicted in Figure 3.4.

To coordinate that process, the team can leverage Integrating Events (shown as purple diamonds in Figure 3.4) that are scheduled using Decision-Based Scheduling — both key enablers that will be discussed later in this chapter. Ultimately, the goal is to close just enough of the knowledge gaps that the team can establish "Success is Assured" prior to making those key decisions that would be costly to change later. And we want that whole Decision-Focused Learning process to be completed very early in the larger development process, certainly before detailed design begins.

The ability to do that better and faster can become a huge competitive advantage. So, organizations should be working hard to continuously improve that ability. Fortunately, for the same reasons that Set-Based knowledge can be effective for collaboration across areas of expertise, it can also be effective for knowledge reuse across projects. The enablers for that will also be discussed later in this chapter. But the net result should be the larger Learning-First Product Development process depicted in Figure 3.5.

The Decision-Focused Learning process is not only feeding the design decisions that must be made in the *Product Value Streams* but also feeding the growing body of Set-Based knowledge that makes up the *Knowledge Value Stream.*

Key Enablers for Achieving "True North"

Earlier, we discussed all the ways that Set-Based helps reduce the workload and increase focus and efficiency, thereby accelerating the learning. However, Set-Based is just one of the enablers of Decision-Focused Learning that in turn enables "True North." In the remainder of this chapter, we will discuss those

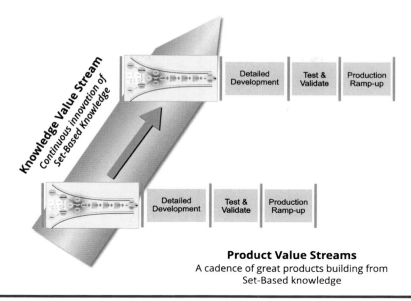

Figure 3.5   Learning-First Product Development process.

additional Enablers, and in doing so will give more clarity to the nature of the Set-Based practices that are infused through the following:

- *Look, Ask, Model, Discuss, Act (LAMDA)*: Enables a culture of highly efficient collaborative learning, accelerating learning at the lowest level while ensuring it remains connected to the bigger picture, including all perspectives.
- *K-Briefs*: Enable the collection and organization of the visual models that facilitate the LAMDA Discussions to solve a problem or formulate a design, further increasing efficiency and maintaining proper focus.
- *Decision-Based Scheduling*: Larger milestone dates pull when the key decisions need to be made. Knowledge gaps must be closed to make these decisions. This defines the minimum knowledge that must be learned and by when.
- *Integrating Events*: Enable a consensus decision-making process that integrates the team around the decisions that pull the knowledge that establishes that "Success is Assured."
- *Mentoring and Cadence*: Enable an efficient process of ongoing guidance and replanning as the learning rapidly unfolds and continuously narrows the design space, allowing refined focus.
- *Trade-Off Charts*: Provide visibility to the sensitivities that allow identification of the better and weaker parts of the multi-dimensional design space, and thus enable deeper understanding and innovation while allowing continued refinement of the focus.
- *Knowledge Reuse and Continuous Improvement*: Enable effective reuse and continuous improvement of existing knowledge, minimizing the relearning required on subsequent decision-making efforts, not just in future projects but in this project (decisions are being steadily converged).

■ *Causal Mapping and Decision Mapping*: Enable most of the preceding enablers. These make visible the causal relationships between the decisions and the targets. They break down the complexity into its individual relationships, greatly simplifying each element that needs to be learned. In doing so, they may have as big a focusing (workload-reducing) effect as Set-Based. These are so important that they are the focus of Chapter 4.

In addition to their individual benefits, these enablers can be a powerful mechanism for making the shift from point-based to Set-Based behaviors. **Applied together, Set-Based and the other enablers represent a fundamental paradigm shift in how the front end of product development is done, accelerating the learning by order of magnitude, enabling "True North" to be achieved in the earliest phases.**

LAMDA for Collaborative Learning

Allen Ward introduced the acronym LAMDA to describe how Toyota engineers did their collaborative work. He observed how they consistently followed the same learning process at every level and for every problem.[7] We like it because it makes it easy to remember the things your team should be doing in any learning situation. If you are in a meeting and there is disagreement, which indicates that a knowledge gap exists, we want you to think, "Which part of LAMDA should we be doing now?"

■ **Look** for yourselves, with an open mind, to make sure you are not missing key facts that would otherwise be missed.
■ **Ask** questions to gain deeper understanding from others' perspectives. Not just "Why?" five or more times; "So what?" and "How?" may need to be repeated to get to true understanding.
■ **Model** visually what you think you know for efficient communication and richer collaboration; without visual models, the pictures in people's heads during a discussion will almost always be different in important ways.
■ **Discuss** with all who have expertise and with all stakeholders who may be impacted to make sure all issues are identified, all opportunities for innovation are explored, and consensus decisions are formed.
■ **Act** on that learning such that the knowledge can be tested and validated as quickly as possible (such that you can *look* at the results and incorporate that into your ongoing learning).

LAMDA enables deep collaborative learning, critical thinking, and consensus building, which are all critical for complex development processes and valuable for basic problem-solving. For LAMDA to be transformative to an organization,

---

[7] Allen C. Ward, and Durward K. Sobek II, *Lean Product and Process Development*, 2nd Edition, Cambridge, MA: Lean Enterprise Institute, 2014, p. 75.

Figure 3.6   LAMDA behaviors.

not just effective for small teams, leadership must embrace its philosophy and ensure its consistent application.

Simply adopting LAMDA can represent a culture change for an organization. It alone can transform the nature of all meetings, conversations, and interactions — most of what we do in product development.

Note that LAMDA is not a linear series of steps. Asking very often leads to more Looking. While Looking, you may be building a visual Model of what you see. Discussion is usually around a visual Model, which should be modified on the fly as knowledge is shared to maximize the efficiency of that Discussion. Discussions should be predominately people Asking questions, often about how they can go Look.

Rather than a series of steps, LAMDA should be thought of as five things you should be doing as much as possible throughout any collaborative learning or analysis situation. For that reason, we depict LAMDA as concurrent behaviors, as shown in Figure 3.6.

LAMDA is often mistaken as a variant of the Plan-Do-Check-Act (PDCA) Deming Cycle. First, LAMDA is not a cycle. Second, and more importantly, LAMDA should occur repeatedly, many times over, during just the Plan phase of PDCA. LAMDA will likely occur in each of the other phases as well.

Similarly, LAMDA should occur repeatedly in each step of whatever Problem-Solving process you use. LAMDA is not an alternative Problem-Solving process; rather, it is an enhancement to any problem-solving (or any collaborative learning) process.

K-Briefs for Organizing the Visual Models

In order to efficiently conduct LAMDA Discussions with as many experts and stakeholders as are valuable, you need to be very efficient at presenting the visual Models that support efficient Discussion. Each time you have such a Discussion, you will learn how to improve that presentation and those visual Models. To enable that, you need a presentation tool optimized for learning situations — for rapid, ongoing improvement of those visual models and the storyline that connects them.

That is the role of the so-called A3 used by Toyota,[8] named for the size of paper on which many saw their Problem A3s. However, not every collection of visual models at Toyota is on an A3 sized sheet of paper (roughly two letter-sized sheets); and the size of the page is not what is important. For that reason, we call it a *K-Brief* (or *Knowledge Brief*); it has the same number of syllables as and even sounds similar to "A3," but it communicates what is important: that it is the carrier of the Knowledge and it is Brief (i.e., as concise as possible).

So, how big should a K-Brief be, if not limited to an A3? That depends on the importance and the complexity of the LAMDA Discussion you need to conduct. If you can only get 10 minutes of an expert's time, you should not bring in a 10-page report to discuss. They won't have time to read it, and thus you'll be having that discussion without visual models; it won't be a LAMDA Discussion. Conversely, if you have scheduled an all-day meeting to work through the complex interactions of a multi-dimensional trade-off in a jet engine, you may need a set of visual models covering an entire wall to ensure all the issues are visible to everybody at once so that they can be worked through.

Some will read the previous paragraph and think, "Ah, thank you, I don't have to try to squeeze this onto one A3 page." In fact, in many cases we may be challenging you to streamline it even more. If you need to communicate the key issues to your team during a portion of the 30-minute stand-up tomorrow morning, such that you may have only 10 minutes to present your content, you may need to be targeting an A4-sized sheet (letter-sized) or less. To maximize the quality of the LAMDA Discussion, you want to optimize your visual models; the more they communicate in the least amount of time, the better.

More than that, in trying to optimize those visual models, you often optimize your thinking — your understanding of the problem. When Toyota mentors push back on a multi-page problem report, they don't say, "You can reformat this to fit on one page," they say something like, "You don't understand this well enough" or "You need to think about this more deeply." Don't make it about squeezing or formatting, make it about synthesizing and distilling the information down to the key things that the audience needs to know, optimizing the visual models and the associated mental models to support the LAMDA Discussion.[9]

The visual Models (the "M" in LAMDA) extend our brains. We can absorb a lot more knowledge and keep that knowledge cached longer as we can continue to see it. A 40-slide presentation will not do that; you need a visual model that can connect the different bits of knowledge together visually to show the interactions, not just describe them.

---

[8] Durward K. Sobek II and Art Smalley, *Understanding A3 Thinking: A Critical Component of Toyota's PDCA Management System*, Boca Raton, FL: CRC Press, 2008.

[9] Durward K. Sobek II and Art Smalley, *Understanding A3 Thinking, A Critical Component of Toyota's PDCA Management System*, Boca Raton, FL: CRC Press, 2008, pp. 15–16 and 112.

So, ideally, K-Briefs are not just documents that collect visual models; they are also designed to make all the visual models visible simultaneously and work together to form a deeper understanding for all participants in the LAMDA Discussion.

And further, the ideal tool for a K-Brief is something that can be easily modified on the fly during the LAMDA Discussion. If somebody says, "Hey, that's not quite right; we need to change the model such that [description of various modifications]," then you do not want the response to be, "Oh, yeah, you're right; I'll make that change before we meet again." Because invariably what will happen is that the next time they get together, that person will say, "Hey, I thought we agreed we'd change that model to show [whatever]." And the response will be, "I did change it… See, I [whatever was done]." "Oh, no, that's not what I meant!"

Whenever you are talking and not drawing, you are probably miscommunicating to at least some portion of the audience. You have some picture in your head, you try to describe it, and they try to understand it and formulate a picture in their mind, but based on how they are thinking about things, not how you are thinking about things, and with their expertise, not with your expertise. What is the probability that the picture in their mind matches yours?

One of the behaviors that Allen Ward observed at Toyota was that they wouldn't be talking for more than a few minutes before they would start drawing.

So, it is important that the K-Brief tool that you use is designed for efficient ongoing modification. For that reason, many people advocate using paper: A3s or larger (flip charts), often pinned to the walls of a conference room, such that there is room to grow, and perhaps augmented with sticky notes to make reorganization quick and easy. However, there are tools that are actually even more efficient at modification than paper, sticky notes, or even whiteboards, and where there are no edges of the sheet to run into. We often see groups conducting an efficient LAMDA Discussion until they hit the edges of their flip chart, at which point they continue talking without drawing and then the discussion soon stalls; they stop making progress. We often let that run for 15 minutes or so before we walk over and point out that they have stopped making progress, and suggest they stick the page up on a wall and blank pages on either side of it, so that they can start drawing again.

The K-Brief templates that your organization uses can form valuable reusable best practices for different situations. For example, your organization's Problem-Solving process can be captured as a K-Brief template such that it actually pulls the process, not only enabling rich LAMDA Discussions underneath that process but pulling those LAMDA Discussions. Similarly, K-Brief templates tailored for analyzing and capturing customer interests have proven effective at getting cross-functional teams to ask the right questions and visualizing the right content to establish a much deeper understanding of true customer interests and the trade-offs that will need to be made.

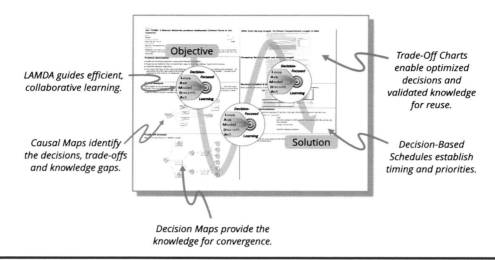

LAMDA guides efficient, collaborative learning.

Causal Maps identify the decisions, trade-offs and knowledge gaps.

Trade-Off Charts enable optimized decisions and validated knowledge for reuse.

Decision-Based Schedules establish timing and priorities.

Decision Maps provide the knowledge for convergence.

Figure 3.7   K-Briefs tell a visual story for collaboration and knowledge capture.

As we discuss the various other enablers in this chapter, consider how they can be organized into a K-Brief presenting the larger storyline, and how a K-Brief template can pull the right collection of visual Models to optimize your LAMDA Discussions. This is depicted in Figure 3.7.

Finally, we should note the two most common failure modes:

- Thinking of a K-Brief as a report, particularly as a report that is written to document what was done (often in compliance with a requirement to have a K-Brief). A K-Brief should be thought of as a tool that pulls and facilitates the process, not as an after-action report. We'll discuss more on that when we get to the Mentoring and Cadence enablers.
- Spending too much time squeezing the content to fit on an A3 page, such that the optimized layout becomes inertia against change. ("Ugh, I worked so hard to get this to fit, I don't want to add that here; let's capture that in a different K-Brief.") K-Briefs are first and foremost ongoing learning tools; while we want to continuously improve how efficiently they present the knowledge, we never want that to get in the way of additional learning. That's why we prefer to leave the fitting of the content onto the screen or paper to the K-Brief software tools; just let them automatically adjust to whatever screen or page you have at the time. Keep the people focused on optimizing the content, clarifying the thinking, and streamlining the communication.

Decision-Based Scheduling

Many reading this will say, "We already do Decision-Based Scheduling; our key milestones or gates are tied to key decision points." However, there are two things missing:

1. In most organizations, the pressure to hit those milestones exceeds the pressure to make knowledge-validated decisions, such that those decisions are made at those milestones even if there are major knowledge gaps. That makes it *schedule-based decision-making*!

2. To pass the milestones or gates, rather than requiring adequate knowledge to make the decisions, there's generally a list of documents that must be created; and to deliver those documents, there's a long series of tasks that must be performed. It is those tasks that drive most of the scheduling. That makes it *task-based scheduling* or *deliverable-based scheduling*, depending on the focus (i.e., do you measure progress by the quantity of completed drawings or completed tasks?).

Task-based scheduling may work for highly repetitive and consistent processes such as manufacturing, where you can optimize the steps of the process to consistently deliver the same quality parts over and over. With that process optimized for such delivery, you can just focus on the execution of the tasks. But in product development or any other process where the work is different every time, particularly where you are looking for innovation (i.e., differences from last time), you cannot rely on the execution of the tasks to deliver the desired result. You must turn the focus onto the result.

Product development doesn't really deliver the product into customers' hands; manufacturing does that. Product development makes decisions about the design of the product and the process of building, maintaining, servicing, and disposing of that product. Its focus should be on those decisions — on ensuring that those decisions will result in a product that is built, maintained, serviced, and disposed successfully. Hence, the focus on "True North" — and a switch to Decision-Based Scheduling.

What does Decision-Based Scheduling look like? Let's start with Figure 3.8.

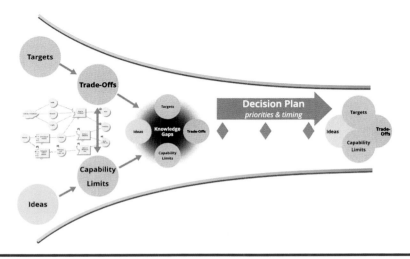

Figure 3.8   Laying out the decision plan for Decision-Focused Learning.

It begins with the active identification of the key decisions that need to be made, starting with the team's ideas on how to achieve the Targets (to what degree we will satisfy the customer interests/requirements), and then the identification of the knowledge gaps that prevent making those decisions today, as delivered by the prior enabler. It then uses the connectivity between those and the innovations on how to close those knowledge gaps to work out a reasonable sequence for making (by which we mean *narrowing*) those decisions. That sequence gets laid out as a rough, high-level schedule against which early progress is measured.

The timing of those key decisions pulls the knowledge gaps that need to be closed, which then pulls the tasks that need to be performed to close those knowledge gaps. But as those knowledge gaps are closed (as learning happens), you will be enabled to narrow the decisions by eliminating the weak parts of the design space. As you do so, some of the knowledge gaps may become irrelevant. So, what knowledge gap to close next and the precise sequence of the decision-making may change. After all, it is a learning process!

The result is a series of LAMDA sprints (as shown in Figure 3.9) to rapidly learn the minimum needed to converge some set of decisions, which typically involves innovating the quickest ways to close the knowledge gaps that are preventing the identification of the weaker parts of the design space (particularly the infeasible parts of the design space). In doing so, risk management activities are replaced by risk elimination activities, and design optimization via "pick the best" is replaced by optimization via "eliminate the weak" convergence.

Note that if you actively move all the learning to the front of the development process, ahead of all the key decisions, then the back end of the process will have much less learning (ideally near-zero), perhaps allowing a shift to more deliverable-based scheduling or task-based scheduling in the back end. That opens up a much larger topic that will be covered in Chapter 5, where we extend the discussion of Decision-Based Scheduling to cover cross-organizational issues.

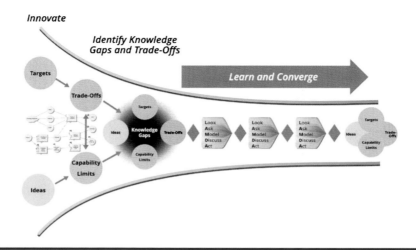

Figure 3.9   LAMDA sprints converge to each decision point or Integrating Event.

Integrating Events

Note the purple diamonds in Figures 3.8 and 3.9, which denote where key decisions need to be made (by which we mean "narrowed") and thus where the knowledge to make those decisions needs to be available. Those purple diamonds pull the learning process. But done right, they can actually do a lot more than that.

By introducing what we call *Integrating Events* at each of those decision points, you can use those not only to verify that you have a consensus decision because you have established that "Success is Assured," but also to establish broad ownership of those decisions across the whole team.

Integrating Events pull together all the key decision-makers and stakeholders and presents to them all the knowledge that establishes that "Success is Assured." If the set of decisions are made this way, it allows them all to look each other in the faces as they give a consensus "thumbs-up" or "go-ahead." In doing so, the whole team is now on the hook for that decision; they all feel ownership. So, if it turns out to be wrong, the whole team feels responsibility to help fix it or otherwise "make it right."

In that sense, what the Integrating Events are integrating is the team!

And knowing what is going to happen at the Integrating Event, the team is motivated to make sure the knowledge is aligned with the decisions prior to the Integrating Event. So, the Integrating Event becomes a very powerful pull of the knowledge needed for the decision-making — not just in the academic sense but in the sense that "I am going to be asked to commit to this at the Integrating Event; what do I want to see before I am willing to do that?" So, it tends to pull higher-quality knowledge, better suited to drive the decision-making.

Many of our clients have pointed to Integrating Events as the most powerful of the enablers. And you heard the important role it played in the Pratt & Whitney story in Chapter 1; hopefully, you now have a deeper appreciation for that. You also saw it as the diamond pulling everything in Figure 3.10 (introduced in Chapter 2), central to the interaction between the three key roles in the Toyota process: the Chief Engineer, the Functional Manager, and the Problem-Solver.

Chapter 5 will give details on how Integrating Events are scheduled along with the related issues regarding managing projects that cross organizational boundaries.

Mentoring

The shift to a focus on the knowledge gaps feeding the key decisions requires engineering managers and technical leads to be more engaged in the knowledge

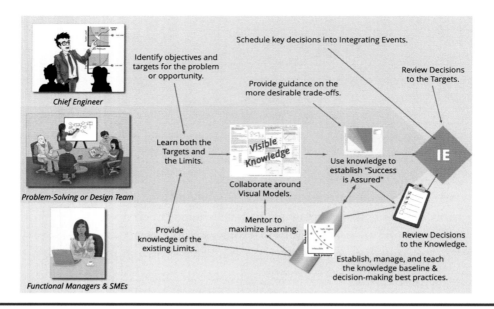

Figure 3.10   Key roles and interactions leading to an Integrating Event.

being actively learned. They need to evaluate the impact of the new knowledge on the next decision and learning activities and make adjustments as necessary. Evaluating that alone would be contrary to LAMDA thinking. But reviewing each bit of knowledge with everyone each step of the way would be inefficient.

The solution is for each K-Brief to have not only a Lead responsible for leading that LAMDA Discussion, but also an assigned mentor responsible for giving guidance on when the learning warrants broader Discussion, Looking, Asking, Modeling, and so on. By having a designated Mentor, that person will be constantly up to date on the content, so re-reviewing does not require re-reading the K-Brief, rather just reviewing what has changed — via quick five-minute updates — maintaining efficiency.

The Mentor will often be a technical expert or someone with more experience and broader visibility, such that they bring a different perspective and can help make the connections that the Lead might otherwise not know to make. The Mentor can truly play an important role to the Lead, helping develop them as an engineer (or knowledge worker) while also helping develop the K-Brief. That process is shown in Figure 3.11.

Many companies created functional leader positions to copy Toyota's leadership structure. Engineers with deep technical knowledge in critical areas of the product were assigned to create how-to documents of the best and/or required design methodologies and limits and mentor these practices to other engineers. What these companies didn't do is assign a functional leader to each of the teams working to close knowledge gaps. Instead, the functional leaders were resources for the team at the team's request. So, if the team leader didn't seek out the functional leaders, the functional leaders were not involved; they were at the gate reviews to review teams' results. Functional leaders regularly questioned the teams' results and required additional work before the phase

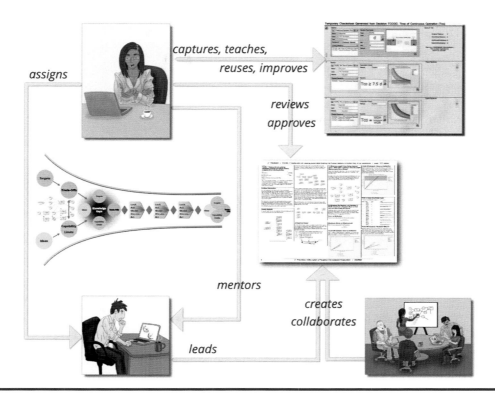

Figure 3.11   Mentored K-Brief development process.

gate could be passed. This made the functional leaders appear as roadblocks to progress instead of valuable resources. Nothing improved with the addition of functional leaders and companies stopped supporting the position. By having the K-Brief Mentor role predominately filled by such functional leaders, this pattern can be reversed and the functional leaders can be fully leveraged and fully appreciated.

Cadence

Note that the shift to Decision-Focused Learning leads to the need for a rapid cadence of engagement. The traditional meeting-heavy schedules of managers, where their calendars can be full for the next week or two, can get in the way of rapid steady convergence and refocusing. Therefore, establishing a mechanism for maintaining a rapid cadence of reviewing and rescheduling becomes a critical enabler of the Decision-Focused Learning that is in turn enabling "Success is Assured."

Such a simple mechanism is to block off an hour a day on your calendar for *mentoring*. Then, on Tuesday, when you meet with one of the K-Brief Leads that you are mentoring, you review the changes to the K-Brief for five minutes, suggest they go discuss the matter with *X* and go Look at *Y*, agree they can get it done that afternoon, and schedule them to drop by to review the learning the next day — which would be difficult to schedule, except that you have an hour-long Mentoring slot on your calendar where you can give them five minutes.

Another effective mechanism is the *Daily Stand-Up*: a short (15–30 minute) meeting that is on everybody's calendar every day, where they review the learning from the prior day with everybody on the team, everybody can give quick feedback on the implications on the decisions or the larger learning, and everybody can suggest ideas for who to discuss the learning with, where to go look, what to Model, and so on. (These need not be every day; perhaps start with just Monday/Wednesday/Friday or Monday/Thursday stand-ups. If you start running out of time consistently, then add more days; adjust dynamically to find the right balance.)

These and other mechanisms for supporting the rapid cadence of mentoring is a critical enabler of not only Decision-Focused Learning but also of the consistent application of LAMDA and ultimately of establishing "Success is Assured."

Warning: Most of us have been programmed to resist high-cadence mentoring! Throughout most of our schooling, we have been given assignments, we work them until they are of high quality, and then we turn them into the teacher to be graded. We do not want to turn it in until it will get the grade that we want. And that pattern is generally carried into the workplace, where our boss gives us an assignment and we don't want to show it to the boss until it is done — full quality.

However, in real-world scenarios, that results in tremendous inefficiency and frustration. Real-world assignments haven't been repeated over and over by past students, aren't carefully refined to be unambiguous, and don't have a singular answer covered by the preceding lessons. Often, much learning is needed even to get the problem statement right. So, if an engineer (or knowledge worker) takes the first cut at that problem statement and runs with it to the final polished solution, then when they "turn it in" to their boss, they are often very disappointed to learn that they've solved the wrong problem or chosen an approach that is not acceptable for some reason that was not stated in the original problem statement. Not only is that inefficient (as depicted in Figure 3.12), but it is also very frustrating and demoralizing to the engineer.

Rather than a process that allows the worker to get far off-track before receiving feedback, a high-cadence process (as depicted in Figure 3.13) ensures that

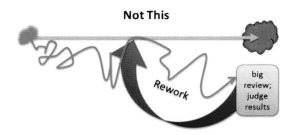

Figure 3.12 Traditional low-cadence feedback-when-done process.

This

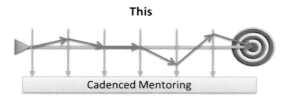

Cadenced Mentoring

Figure 3.13 High-cadence mentoring keeps things on track.

you never get very far off-track. Such cadence results in much less waste, much higher efficiency, and much less frustration.

Despite that, it generally requires a concerted effort by the Mentors to get the engineers (or knowledge workers) to embrace this. The tendency is to not show their work until it is of high quality. The Mentors need to explicitly ask to see each next step, to discourage any polishing or refining, to emphasize it is a learning process and that they don't want to risk wasting time doing the wrong thing well. It can always be polished later, after the learning is done. On the flip side, they should not be critical of roughness or test failures, as long as they "failed quickly" and thus learned quickly.

Trade-Off Charts for Set-Based Decision-Making

To allow time for Decision-Focused Learning to close the knowledge gaps prior to making decisions, we often need to delay the decisions. But other dependencies can get in the way, preventing you from delaying those decisions. The solution is to allow decision-making to narrow to a better part of the design space but not to "pick a point." That allows others to proceed knowing the undesirable part of the design space has been eliminated but without prematurely picking the final point.

However, that requires that those others are able to do that analysis with decisions as ranges of values — as sets. Such Set-Based analysis and decision-making is an important skill to enable "Success is Assured" decision-making. And the key enabler of that is analysis based on Trade-Off Charts. But before we explain that, there is another driver of Trade-Off Charts...

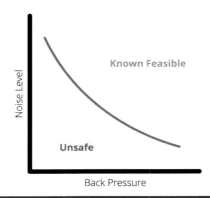

Figure 3.14 Simplified Trade-Off Chart for automobile muffler.

To enable consensus decision-making based on knowledge, it is critical to be able to see the impacts of that knowledge on the decisions. Or more precisely, to see the impacts of narrowing one set of decisions on the other decisions, particularly the customer interest decisions, based on your knowledge of the limits that exist. That is the role of the Trade-Off Chart, as illustrated by the chart for an automobile muffler (exhaust system) shown in Figure 3.14.

Trade-Off Charts are the visual representation of the basic product and process physics and economics. They serve as Toyota's engineers' primary tool to

- Understand the limits of the situation
- Communicate and negotiate between specialties and functions
- Train new engineers
- Negotiate and communicate with suppliers and customers
- Conduct design reviews
- Communicate between developers and managers
- Design quality into the product
- Record/reuse knowledge (Checksheets)

Allen Ward stated that the most important thing he learned at Toyota was their focus and dependence on trade-off curves. It is a knowledge-based mechanism for visually understanding process capabilities and limits, for understanding sensitivities between decisions, for optimizing and converging trade-off decisions, and for creating generalized reusable knowledge.

One of the most impactful things about trade-off curves is that no matter what your expertise is in, you can understand the trade-off curves from others' expertise. You may not have the expertise to know why the curve is what it is, nor how to innovate ways to move the curves, but you can understand the impact of the curve on the decisions that need to be made. Thus, Trade-Off Charts are very effective for collaboration across areas of expertise (Figure 3.15).

Production Support may work out the complex trade-off between *Back Pressure* and *Engine Cost* (to hit other targets such as horsepower to the wheels or acceleration, as Back Pressure increases, a bigger or more expensive engine is needed). Engine Costs above the current maximum Engine Cost target are considered weak. Given they know the trade-off with Back Pressure, they can then communicate that they want to eliminate the weak in terms of Back Pressure, which then allows the designers working on the *exhaust system* to eliminate that portion of the design space. They also eliminate as weak the design space above their maximum Noise target and eliminate as *unsafe* the region below the curve that is likely infeasible. What remains unshaded is the current solution space under consideration. As more knowledge connected to Back Pressure or Noise Level is learned, that space may be further narrowed. The Project Team manages these interconnections between different areas of expertise (what we call *Interface Decisions*), ensuring that the learning is propagated to those who need to know.

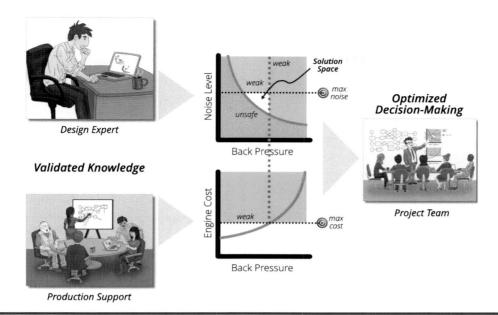

Figure 3.15 Optimized decision-making by combining Set-Based knowledge.

For Trade-Off Charts to be an effective enabler, it is important that the decision-makers can trust them. For that, it is important that they can understand the knowledge behind them and see the source. Further, it is important that they have experts they trust challenge the knowledge. The best enabler for that is a visual model of the knowledge on which the Trade-Off Chart was based. The next chapter will introduce such a visual model: the Decision Map. The Decision Map can also provide the connections the Project Team needs to see to propagate the learning and the ongoing converging of decisions.

Another key challenge in complex product development is the fact that our brains can only see three dimensions, but our trade-off space is almost always many more than three dimensions. Thus, each Trade-Off Chart can only show us a three-dimensional slice through the multi-dimensional trade-off space. To understand that larger space then, we need multiple three-dimensional slices: multiple related Trade-Off Charts. Keeping track of how those multiple Trade-Off Charts relate is a key enabler for complex decision-making. Again, the Decision Map is our enabler of that, and demonstrating how it does that will be the topic of the next chapter and Part II of this book.

Thus, for complex product development, the Trade-Off Chart is the key enabler of Set-Based, which is in turn the key enabler of establishing "Success is Assured." If your teams cannot see the feasible sets that make up their multi-dimensional design space, then it will be nearly impossible for them to operate Set-Based. And if they cannot operate Set-Based, then it becomes nearly impossible to delay the decision-making by converging rather than picking a point, and it becomes difficult to accelerate the learning to feed those decisions.

Building the generalized knowledge to populate Trade-Off Charts takes time. You can't just test one point; you need to test enough points so that you can characterize the entire set. Further, devising the right set of two- and three-dimensional Trade-Off Charts to help you make multi-dimensional trade-off decisions takes time. But there is always pressure on the front end of projects to make decisions quickly. And there is tremendous profitability for those organizations that can reduce their product development cycle times. The time from identification of customer interest to putting a solution into their hands is the key driver of profitability and market share.

So, a critical enabler of "Success is Assured" in the face of competition is *Knowledge Reuse* and *Continuous Improvement* of that knowledge that the teams use to make those key decisions.

Knowledge Reuse and Continuous Improvement

While it is satisfying to see organizations migrate to the use of K-Briefs (or A3s) as their problem-solving methodology, it is equally frustrating to see those same organizations adopt a knowledge management strategy built around the storage and retrieval of those K-Briefs or A3s. In contrast, Toyota isn't concerned with the storage and retrieval of Problem A3s; those A3s finished serving their purpose when the problem was solved. It would be extremely inefficient to reuse knowledge by re-reading the problem-solving stories of every problem ever solved in a company.

The reusable knowledge is not the specific problems solved or the specific solutions that were adopted; rather, it is the underlying knowledge that drove the decisions that led to those solutions. And most often, that underlying knowledge is in the collection of Trade-Off Charts that define the design space. At Toyota, those are captured into their Engineering Checksheets that their functional leaders own, teach, and continuously improve. Whatever tools your organization uses to make its decisions, that's where your reusable knowledge needs to be captured.

Knowledge is not truly learned by an organization unless it has become integral to the organization's decision-making behavior. Knowledge captured in a repository, no matter how valuable and how easily searched, is not really learned if it does not consistently impact the organization's decision-making.

A corollary of this is that if an organization is not using Trade-Off Charts (or an equivalent knowledge-based tool) to make decisions, then its knowledge has no place to be captured in a reusable way, and organizational learning will be difficult, if not impossible.

Fortunately, the same properties that make Trade-Off Charts very effective for collaboration across areas of expertise and for driving optimal decision-making also make Trade-Off Charts very effective for reuse across successive projects.

Consider the prior example of a Trade-Off Chart for an exhaust system muffler where the team wants to minimize the noise but at the same time minimize the Back Pressure on the engine (since that reduces horsepower to the wheels). Imagine when developing the muffler for the next Camry, the team experimented, made modifications, and after a first failed prototype arrived at a good

point design (as shown in Figure 3.16). In that case, they will seem to have succeeded as they will indeed have a better muffler. But what have they learned for
the next product development effort?

Well, they've learned that they can't go as far over as Prototype 1. So, based
on that, on the next Camry project, they might shoot for those noise reductions
but not so much Back Pressure reduction. And with a little rework (another
failed prototype), they end up at a further improvement on that next Camry (as
shown in Figure 3.17). They have again made improvements, but what have they
learned? It could be that you just cannot reduce Noise any lower, as shown in
Figure 3.18a. Or it could be that you just cannot reduce Back Pressure any lower,
as shown in Figure 3.18b. Of course, most likely it is somewhere in between, as

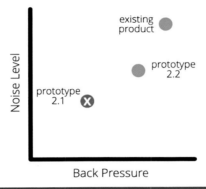

Figure 3.16   Developing the second generation of a muffler.

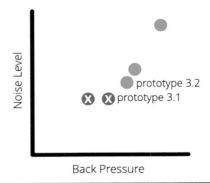

Figure 3.17   Developing the third generation of a muffler.

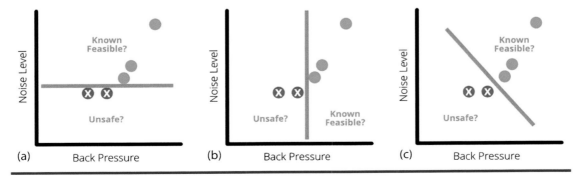

Figure 3.18   Given those points, what have we learned?

shown in Figure 3.18c. They have learned very little other than that one point in the design space.

If their goal is to design another Camry with the same muffler performance, then that limited learning may be okay. But what if their goal is to design a Lexus sedan where they want much less Noise but are willing to have higher Back Pressure since they can afford to have a larger engine? Or what if their goal is to design a sporty Scion hatchback where they want to minimize Back Pressure on the small engine, but a little more Noise is okay (it's a sporty car, so customers want to hear the engine)? In those situations (Figure 3.19), they essentially know nothing, although they've been designing Camry mufflers for years.

Toyota would have wanted to understand the trade-off curve from the very beginning. So, on the first Camry effort where they realized they lacked that knowledge, they'd have done the testing required to establish that curve (Figure 3.20). They would not have just tested in the immediate vicinity of the current point design; they would have tested across the whole of the design space such that they established a broader understanding of that trade-off. Then they would have been able to reuse it in all of their design work on future cars.

That is what we mean when we refer to Set-Based. We might test and measure specific points, but it is with the goal of understanding the infinite set

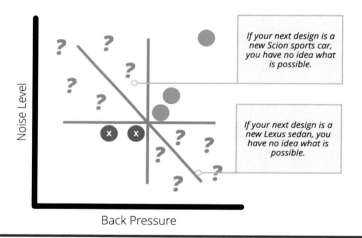

Figure 3.19 We have learned very little other than that one point design.

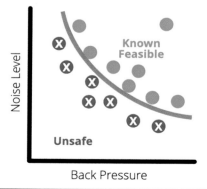

Figure 3.20 Test-before-design gives clarity on the design space (infinite points).

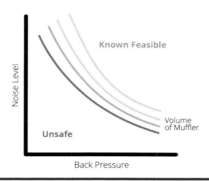

Figure 3.21   Adding a third dimension to the Trade-Off Chart (volume).

of possibilities (anything above and to the right of that curve is possible) by establishing visibility to the boundaries of the design space where "Success is Assured." (Recall the Wright Brothers: they similarly collected wing data across the entire design space, creating a comprehensive understanding that they then were able to reuse when designing their highly efficient propeller.)

Of course, this example is oversimplified. The reality is that there is a third dimension to this trade-off: designers can increase the volume of the muffler to reduce the Noise without impacting the Back Pressure (Figure 3.21, where the brown line at the top is the smallest volume and the blue line at the bottom is the largest volume). But then they run into limits of what will fit on the car, which may vary depending on the car design.

And then there is a fourth dimension: weight. A fifth: the materials used. A sixth: the cost. And so on. Building up that multi-dimensional knowledge takes time and effort. However, experience has shown that such time and effort is much less than initially feared. That is where the Causal Mapping methodology comes into play.

Learning the Reusable Set-Based Knowledge

Causal Mapping (as will be described in the next section) breaks down complex relationships into individual relations that can be more easily learned — more easily tested across the design space. Typically, the vast majority of those pieces are well known, and the few that are not have been adequately simplified by the breakdown to make it relatively straightforward to acquire the required data.

On that point, note that the team is not analyzing or testing point designs for the whole car! Or even for the whole exhaust system. That would incur much of the inefficiency of point-based design. Rather, they are just analyzing or testing the much simpler relations that were identified by the Causal Mapping step. For those individual relations, the infinite set for each can typically be characterized quickly. Then those infinite sets can be put together, as shown by the Causal Map, into a view of the larger design space for the exhaust system or car.

This is a good time to emphasize the importance of the "L" in LAMDA. When the team answers, "I am not sure how to analyze the behavior of this particular relationship," often the best question to ask is, "Well, can we just go *Look*? Is there some way to setup a simple measurement on an existing example?" And remember, when asking that question, that you don't necessarily need high precision; you often just need to establish where the limit is — where the worst case is. Too often, teams get bogged down thinking they need an elaborate model that accurately computes the point values for any point design in the space, when often they just need to characterize the range of possible values or even just the worst-case value with relatively low precision and show that such limits will not compromise "Success is Assured." Sometimes adding a little padding or margin into the design can greatly reduce the need for precision in the model or testing.

On the flip side of that point, there is often an *N*-dimensional relation that the team suspects could be broken down further, but the inner workings are not understood or easily discovered. In such cases, consider the "L" in LAMDA: "Can we just go look at the subsystem at that level and take appropriate measurements?" You do not need to break down a relation into its components if you can more easily just measure how it performs at that granularity. Some insight and opportunities for innovation may be lost by not breaking it down further, but if you can determine the relationship, there is no need to see what's inside in order to successfully design around it. Further breakdown can be left as a future exercise of continuous improvement of the model, if the potential improvements justify it.

Returning then to the Problem K-Brief: its role is to lead the team through an effective problem-solving process, identifying the faulty or missing knowledge that led to the wrong decisions. The new or improved knowledge is then captured into the organization's decision-making tools so that they can be used to decide on the right solution to the problem. The Problem K-Brief documents both the learning and the decision-making such that it can be reviewed, approved, implemented, and checked; but the reusable knowledge is the knowledge that was captured into the decision-making tools that represent the organization's best practices for making those decisions, both now and in the future. In other words, the Problem K-Brief is the mechanism for learning and problem-solving, not the mechanism for knowledge reuse. That is the role of the Decision Map that evolves from the Causal Map.

Enabler of the Enablers: Causal Mapping

The enablers discussed above call for

- Visual models that can facilitate LAMDA Discussion across people with a wide variety of different expertise
- Models that visually connect the Ideas and the Capability Limits to the Targets you are trying to achieve, such that the Trade-Offs become visible
- Models that help you focus on the key Decisions to be made
- Models that help you identify the Knowledge Gaps that need to be closed to make those Decisions

- Models that help you see the existence of unknown unknowns (that something is missing, even if you don't know what is missing yet)
- Models that help you see the dependencies and thus the priorities when doing Decision-Based Scheduling
- Models that help you break down complex relationships into their simpler component relationships that can be more easily measured or otherwise characterized.
- Models that can help you formulate the Trade-Off Charts that you use to see the design space (the sets of alternatives), to optimize your decision-making, and to build consensus decisions

There are many visual models that can be helpful, but the one visual model that we have engineered to excel at all of that is the Causal Map. Our Causal Mapping methodology starts with a simple series of questions designed to identify the key cause-and-effect elements that need to be mapped together in order to better understand the problem and/or design space that needs to be tackled by the team. As depicted in Figure 3.22, it is connecting the Targets you are trying to achieve, and the Ideas you have for achieving those, to the fundamental Capability Limits that you must respect, such that you can characterize the Trade Offs that must be made in the design of the solution.

Our Causal Mapping methodology then continues asking questions that help evolve that Causal Map into a more rigorous Decision Map that can serve as a visual representation of your team's best practice for making the set of decisions that need to be made (Figure 3.23). The Decision Map visually defines the *structure* of your multi-dimensional design space. (The associated Trade-Off Charts visually show you the *limits* of that design space.)

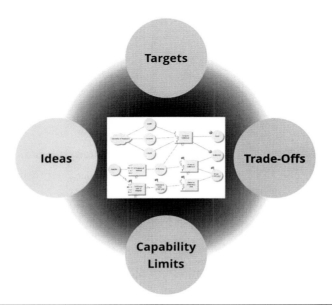

Figure 3.22 Causal Maps connect what you know to reveal what you don't.

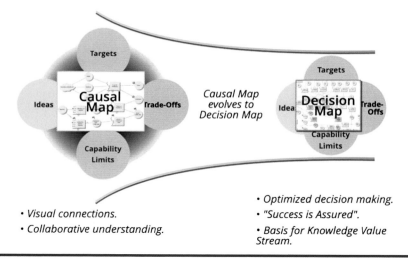

Figure 3.23 Causal Maps and Decision Maps play important roles.

The first time through, the Decision Map is the visual representation of the actual knowledge being used to make the decisions on this project. But as more is learned and the decisions need to be further refined, that Decision Map will continue to evolve to represent how those decisions should be made in light of the latest learning. Given that, the Decision Map then becomes the vehicle for continuous improvement of how an organization makes a set of decisions on all future projects, evolving with the learning on each of those projects. The Decision Map serves as the core of the knowledge flow from project to project.

As mentioned in Chapter 2, a key lesson from Toyota is that product development is not about developing a series of individual car models; instead, it is about developing an ongoing stream of knowledge that results in a cadence of great cars entering the marketplace. As illustrated in Figure 3.24, product development should be the ongoing development of a knowledge value stream that continuously produces great products. The front end of every new product

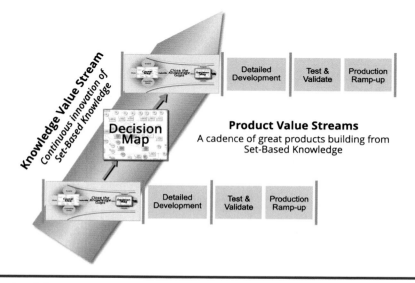

Figure 3.24   Decision Maps provide highly reusable Set-Based knowledge.

development effort should build from existing Set-Based knowledge, evolving that knowledge to meet the new targets for the customer interests. The Decision Map is the vehicle for that ongoing evolution, as it provides the visual connectivity between everything that needs to be considered as new technologies, new approaches, and other innovations are considered to meet those new targets.

To make the roles of the Causal Map and the Decision Map clear, Chapter 4 will use an example of a *material selection* problem. Most engineers inherently know that changing the material of a component might impact its reliability, strength, weight, cost, and other performance characteristics. But what if they are unknowingly affecting other aspects of the part? As product complexity increases, the required collaboration to uncover such issues becomes more difficult, while the likelihood of such interactions becomes higher. The Causal Map serves as a roadmap connecting the right people and focusing them on the cause-and-effect relationships between what they know and what those others know.

To show how the Causal Mapping methodology and the resulting Decision Map can deal with such complexity, we will need a more complex example than the Material Selection problem used in Chapter 4; that is the purpose of Part II of this book, which will show four independent organizations collaborating to solve an aerospace example with real-world complexities.

All of the enablers in this chapter are enabling efficient and effective learning to close knowledge gaps. These gaps are numerous, complex, and interrelated. The enablers all fit together to form a simple methodology of collaborative learning, resolving, and documenting. It is a methodology that is suitable for the earliest phases of product development, when many things are inherently fuzzy, but the most critical of decisions must be made.

As you read the following chapters, think about your current design and knowledge capture processes. Do they give the larger context — the big picture? Do they keep your decision-makers informed on how their decisions will impact that big picture? Can your engineers see the trade-offs that need to be made and why? Can they see the design space that is available? Are they given guidance on efforts that may open up the design space? This is the most important part of the book. Our process is not just passive documentation of past learning, like a lessons-learned database; rather, it is visual models that enable superior decision-making and that can be continuously improved.

Chapter 4

Causal Mapping

We are not the first to use the term *Causal Mapping*, so let us begin by saying this is not the same as any other Causal Mapping that you have seen before, though it may be similar in various ways.

Our Causal Mapping is a methodology that starts with the development of our generalized Causal Map visual model but with the express intent of evolving that into a Decision Map visual model (both Maps will be fully defined later in this chapter). The Decision Map ties the Set-Based knowledge available from the various areas of expertise in your organization to the decisions that you need to make.

The purpose of our Causal Mapping is not simply *root cause analysis* (RCA); rather, its primary purpose is *knowledge-based decision-making*. However, we also use it for RCA because, in the end, RCA is not very valuable if it doesn't help you decide how best to solve the problem. If solving the problem requires changing the design of your product or related processes, then you really need to be considering all the same things the original designers were considering (plus whatever it was they missed that caused the problem). In other words, problem-solving (the typical purpose of RCA) is a special case of decision-making, and it should be knowledge-based decision-making.

In the context of knowledge-based decision-making, the "Causal" in Causal Mapping should not be taken as the "cause of a problem"; rather, it refers to the causal relationships between the design decisions you are making and the customer interests you are trying to satisfy. (Or even more generally, the cause-and-effect relationships between whatever decisions you are making and whatever objectives you are trying to achieve by making those decisions.) **The only way we can properly make decisions is if we understand the effects that making those decisions will cause. That is what Causal Mapping is about**.

Inspiration for Causal and Decision Maps

Although the typical evolution is from general Causal Map to refined Decision Map to Trade-Off Charts showing the sensitivities and the boundaries of the design space, the inspiration and justification for these two closely related Maps came from the opposite direction: Trade-Off Charts inspired the Decision Map, and the Decision Map inspired the more general Causal Map.

So, let's begin this chapter where we began more than a decade ago, with the Decision Map. We will then explain our journey to the need for a generalized Causal Map to help people develop their Decision Map. That then leads to our Casual Mapping methodology for developing a Causal Map and then evolving it into a Decision Map.

Trade-Off Charts Inspired the Decision Map

As discussed in Chapter 3, Trade-Off Charts are a key enabler for Set-Based, which is a key enabler for establishing "Success is Assured" prior to making decisions. Allen Ward observed Toyota's extensive use of Trade-Off Charts to

- Understand what is known
- Communicate that across areas of expertise
- Communicate and negotiate with suppliers
- Conduct design reviews
- Design quality into the product
- Capture knowledge in a way that could be reused and continuously improved

To do that, Toyota Functional Managers collected those extensive sets of Trade-Off Charts into their Engineering Checksheets, which were used for design reviews. The engineers doing the knowledge work leading up to those reviews would of course want to leverage that knowledge. But doing so required the support of those Functional Managers to make clear how all those Trade-Off Charts were related.

When trying to teach such engineering practices to organizations with complex product designs, the Functional Manager's role of managing and mentoring the relationships between all of those Trade-Off Charts proved to be a challenge. Toyota Functional Managers grew into those positions, having been doing knowledge-based decision-making based on those Trade-Off Charts for decades. By that time, they had those relationships in their heads. But we want to ramp up new organizations into these practices in weeks or a few months, not in years. Further, even if an organization was willing to wait for a longer ramp-up, few organizations have engineers consistently staying in roles long enough to achieve it.

So, Targeted Convergence developed **the Decision Map: a visual model for tying together the knowledge (in the form of Trade-Off Charts) to the**

collection of decisions that need to be made based on that knowledge. For those who haven't been making knowledge-based decisions for the last decade, it serves as an effective road map from the decisions you know you need to make to all the knowledge you need to be considering as you do so.

The Decision Map consists of just two different shapes: each known limit or fundamental trade-off relationship is represented as a rectangular *Relation* shape connected to one or more circular *Decision* shapes that represent each axis of that Chart. So, for example, the last muffler example from Chapter 3 would be visually modeled as shown in Figure 4.1.

Note that none of the three decisions are marked as inputs versus outputs. The box does *not* represent an activity or task to be done. Rather, it represents how any two of those decisions limit the third. If you decide you want less *Noise*, then you will either have more *Back Pressure* or you will have more *Volume* (or both). If instead you decide you want less volume, then you will get more Noise or more Back Pressure (or both).

We mentioned in Chapter 3 that those trade-offs may also be related to the *Mass* of the Muffler and/or to the *Material* used in the Muffler. How would that look in the Decision Map? That depends on the nature of those relationships. For example, if the Mass is simply computed from the Volume and the *Material Density* and do not otherwise affect the trade-off between Noise and Back Pressure, then the Decision Map would look like the one shown in Figure 4.2.

Alternatively, if the Mass and Material have independent effects on the Noise versus the Back Pressure, then those all may just come into the one multi-dimensional *Relation*, as shown in Figure 4.3.

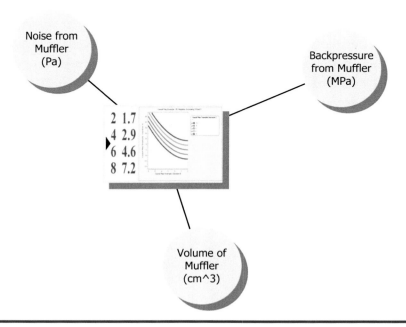

Figure 4.1   Causal Map of the 3D muffler Trade-Off Chart in Figure 3.21.

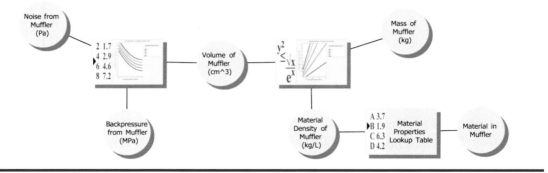

Figure 4.2 Adding more dimensions to the muffler trade-off (Possibility 1).

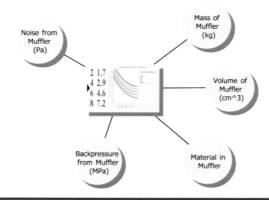

Figure 4.3 Adding more dimensions to the muffler trade-off (Possibility 2).

Either way, the engineers would be able to compute a Trade-Off Chart between those five Decisions; the only difference would be whether they were using three separate relations to compute it or just the one.

Thus, the Decision Map gives the decision-makers clear visibility of all the Relations they need to combine in order to compute the larger trade-offs between any set of decisions. **In other words, the Decision Map provides visibility of the knowledge needed to establish that "Success is Assured" for a set of decisions.** The Decision Map serves as our mechanism for capturing a team's best practices for decision-making in a way that is highly reusable, continuously improvable, and more visually navigable than Toyota's Engineering Checksheets (which were the inspiration for the Decision Map).

Further, when trying to determine the required knowledge for the larger trade-offs, by breaking it down into the various independent pieces that together form all the necessary connections, the effort required in collecting that knowledge is greatly simplified. Very often, those smaller independent pieces are well known. And thus, the Decision Map serves as our tool for identifying and closing the knowledge gaps regarding how the Ideas and Capability Limits tie to the customer interest Targets and the resulting Trade-Offs, as illustrated in both Chapters 1 and 3 (Figure 4.4).

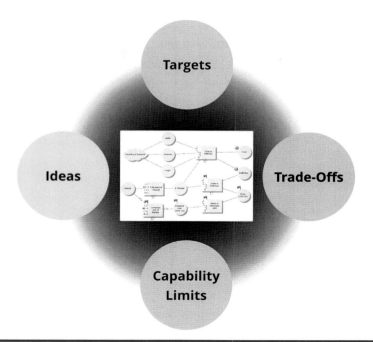

Figure 4.4   Mapping the Targets to the Limits to optimize the Trade-Offs.

The Decision Map Inspired the Causal Map

When teaching organizations to use our Decision Map, we would initially suggest starting from their existing decision-making tools. If they had any existing Trade-Off Charts or existing spreadsheets that computed things for them, converting those into Decision Maps was straightforward (as seen in the previous muffler example).

However, the greater need for the Decision Maps was where they had loop-backs, typically where they had knowledge gaps (not where they had Trade-Off Charts or spreadsheets). For those parts of the Decision Map, we would recommend starting from their RCAs of those problems (i.e., the root causes of the loopbacks), which we allowed them to do the traditional ways they have always done them. We saw no point in teaching RCA as it was a standard part of engineering education and a standard part of most Problem-Solving processes that people already knew.

What we found was that most of the engineers had trouble getting from a blank sheet to a good RCA. While there were certainly successes with most of the RCA tools and methods, that was when the tools were in the hands of people who would have done a good RCA no matter what tool they used. The tools were *not* helping the rest to carry out a good analysis. In fact, they were often getting in the way of doing what needed to be done (e.g., fishbone diagrams make it difficult to go further than three deep, whereas even simple five-why analysis calls for going five deep).

Further, when we looked at many of the enhanced RCA or Root Cause and Corrective Action (RCCA) methodologies companies had adopted, or related Failure

Mode and Effects Analysis (FMEA) variants, they had grown into project management tools or exhaustive analysis tools with tremendous overhead. Rather than being lightweight tools that helped the engineers get their jobs done more quickly, they were actively avoided because of all the excess work that would be necessitated.

Having analyzed the many RCA and Problem-Solving tools and methods in use, almost all of them were found to suffer from several critical flaws:

1. They don't have a good terminating condition. For example, 5-Why Analysis doesn't really suggest you stop at 5: you keep asking "Why?" until you get to the "root cause." But how do you know when that is? Without a good terminating condition, people will tend to be unreliable at getting to the root cause *and* will tend to work too hard digging deeper than necessary in many cases.
2. They tend to cause other problems when fixing the current problem. Often these occur in cycles: fixing problem A, they caused B; fixing B, they caused problem C; fixing C, they re-caused A; fixing A and avoiding B, they caused D; fixing D while avoiding A, they caused C; and so on.
3. They tend to produce sub-optimal conditions or undesirable trade-offs. Or said another way, they don't give good guidance for judging a better solution to the problem versus an inferior solution to the problem.

Most of the methodologies give a lot of pointers for how to analyze data to uncover causes and get to the "root cause." But then the analysis step is deemed complete and they move to brainstorming solutions or even selecting the best solution, as if knowing the root cause is *all* you need to decide on a solution.

In reality, getting to root cause is just one-third of the causal analysis that needs to be done to properly solve a problem (to make the right decisions). Once you've worked down the causal chain(s) to the root cause(s), you then need to work back out again to all other customer interests that may be impacted by changing the root cause(s) – RCA in reverse, you might say. And even with that, you're only two-thirds done. You can't devise good alternatives let alone choose the best solution based only on causal connections; you need to know the sensitivities and, based on those, the trade-offs between the customer interests. Until you can see the trade-off curves, you don't know enough to make the right decisions to best solve the problem. Only then have you completed the causal analysis that is required to truly understand the problem.

In other words, to properly solve a problem, you have to make the right decisions on what to change; essentially, you need a good Decision Map. So, we began developing our own RCA methodology and associated visual tools driven from two sides:

1. The flaws we were seeing in the traditional RCA tools
2. The Decision Map that we were ultimately wanting the causal analysis to grow into, whether for larger product development purposes or for problem-solving

Targeted Convergence's solution, **the Causal Map, is a simple extension of our Decision Map**, which was already designed to handle the complexity, to enable collaboration on that complexity by experts across all domains, and ultimately to be the clear end point that we were trying to get engineering teams to anyway. Thus, the Decision Map provides a very strong terminating condition for Causal Analysis (its traditional weakness). The weakness of the Decision Map was that it was a difficult first step for those new to it.

The Causal Map simply adds two shapes in addition to the circular Decision and rectangular Relation shapes of the Decision Map: a "scroll" shape for more general causal statements that are neither Decision nor Relation and a "cloud" shape for fuzzier statements that are likely a number of different Decisions, Relations, or Causal Statements that have not been broken out yet (or are broken out in a separate Causal Map). By adding those two additional shapes, it becomes easy to capture people's thoughts and ideas before they are well formed in terms of the Decisions that need to be made and the Relations between those Decisions. It gives teams time to think through complex situations more loosely and then evolve parts of that Causal Map gradually into Decision Maps (subsets of the Causal Map that are all Decisions and Relations). It enables rapid learning pulled by the decision-making that needs to be done.

The Causal Map provides an easier starting point for the teams we train; and from there we just needed to develop a highly reliable Causal Mapping methodology that would work even for engineers who would *not* have naturally come up with a good causal analysis anyway. That took some time, but what emerged was a series of steps in the form of simple questions that engineers could keep answering, and an effective Causal Map would consistently emerge, evolving naturally into a Decision Map.

In the rest of this chapter, we will describe our Causal Mapping methodology, which fixes the critical flaws described previously, enables collaborative innovation, facilitates evaluation of the alternative innovations using trade-offs between the competing customer interests, and creates a knowledge document that can be used by future teams facing similar issues.

Given those characteristics, our Causal Mapping methodology is well suited for *all* decision-making scenarios, not just problem-solving. Anywhere there are multiple competing customer interests for which you want to make decisions that optimize those trade-offs, this is the methodology for you. It can be used for

- Designing a new product
- Developing a more profitable manufacturing process
- Improving an organizational process
- Working out a more effective go-to-market strategy
- Optimizing a supply chain or inventory system
- Any other complex decision-making with multiple objectives

The tool allows you to identify the impacts your decisions will have on those objectives so that you can see the trade-offs you are making. Then you can pick the best spots on those trade-off curves. If you can't see where your different decisions will put your results on those trade-off curves, then you are essentially guessing and hoping you get lucky.

But at the same time, our Causal Mapping methodology is also better for "simple" RCA as it covers all three thirds of the analysis needed to properly solve the problem.

- *First third*: Finding the Root Causes (with clear terminating condition)
- *Second third*: Finding all that is Impacted by those Root Causes
- *Third third*: Making Visible the Trade-Offs between those Impacts

We will cover each third later in this chapter; each with an example to illustrate. And then we will show how those three thirds are needed anytime you are changing the product or process design (which you are when you are solving a problem).

Causal Mapping for Problem-Solving

Although it is tempting to teach our Causal Mapping methodology by simply listing the step-by-step questions used in the initial map creation and then demonstrating them with an example, we fear that many may find that a little too dry, particularly if they are not yet convinced they want to adopt our Causal Mapping methodology. So instead, we have put the Causal Map details and the step-by-step Causal Mapping methodologies into Appendices III through VI of this book.

The remainder of this chapter and Part II will teach our Causal Mapping methodologies by allowing you to experience Causal Mapping in the form of two examples. This chapter will use a simple but very common example: material selection to fix some problems with a new product. We hope most organizations will be able to relate it to problems in their own space, making it an effective way to learn our Causal Mapping.

However, most will also object that this example is much simpler than their problems and will question if it really applies to complex problems. To answer that, the second example in Part II is not a simple problem-solving scenario but rather a complex system-of-systems design problem crossing not just areas of expertise or organizational boundaries but also corporate boundaries. We hope that most will find that aircraft design example sufficiently complex that they will have confidence the methodology and visual tool are adequate for their own problems.

But for now, let's focus on the following generic material selection problem, starting with a proper Problem Description.

Problem Description

Although developing a Problem Description is really the first step in the larger Problem-Solving Process, rather than part of the Causal Analysis Process, we would be remiss in skipping this step here because

- The most common failure mode for Causal Analysis efforts is a poor Problem Description. Too small a scope results in not addressing the whole or sub-optimizing; too large a scope results in a slower problem-solving process or even complete stalls as the process becomes overwhelmed; and the wrong focus results in solving a symptom or a lower-priority problem than needed.
- A good Causal Analysis process should be robust in the face of a bad Problem Description. That is, it should help you see that your Problem Description is flawed and lead you to an improved Problem Description. So, the development of the Problem Description is, in a sense, part of the Causal Analysis Process.

Given those two facts, we will start our Causal Analysis Process with its input: the Problem Description.

Whenever possible, a Problem Description should begin with going and looking at the problem with an open mind, making direct observations uninfluenced by premature solutions, collecting facts and data, and creating visual models of what was seen to support collaborative discussion with those impacted by the problem as well as those who might have expertise on the situation. This is an example of where we apply the LAMDA enabler introduced in the last chapter, as it encourages all of those behaviors.

While *Looking*, you are often in the presence of those who really understand (those who live with the problem); so it is a great time to do some LAMDA *Asking*. And if you build the LAMDA Model of what you saw on the fly, then you can review that model with those experts and ask for their feedback on that model – the first LAMDA *Discussion*.

All of that may happen before you even write the first Problem Description, though you typically need some sort of Description so that you know what to go Look at and apply LAMDA to.

In that first round of LAMDA, be sure to ask about the significance of the problem: what does it impact? And what should be the objectives of the Problem-Solving effort? Because, in the end, the Problem Description needs to include objectives.

Table 4.1 summarizes the application of LAMDA to the Problem Description.

One critical test for the Problem Description is to ask, "How do we know when we are done? How do we know that we have solved this Problem?" If that is not clear from the Problem Description, then it needs more work.

Table 4.1   LAMDA Applied to the Problem Description

Look	Go look and see the problem for yourself with an open mind (with no solution or even preconceptions) and collect the facts.
Ask	Ask "What facts do we know or think we know?" Ask "What is the significance? (So what?)" Ask "Is this just part of a larger problem?" Ask "Who are the customers? Who is most impacted?" Ask "How is this impacting the customers?" Ask "What are our measurable objectives?" Ask "How will we know when we've succeeded in solving the problem?" Ask "Is this problem similar to another problem in a different context that has already been solved?"
Model	Model the problem and its significance visually. Show a timeline or photo whenever appropriate. Show how to quantify the impact on the customers.
Discuss	Make sure that all understand and agree with the facts, the significance, and the objective (the criteria for declaring "success").
Act	Quickly act on this learning by doing a Causal Analysis on what is required to satisfy the objectives, thereby revealing missing objectives and causal drivers.

Note the *Act* for the Problem Description. LAMDA Act is about putting the learning into action so that you can test its correctness sooner rather than later. Causal Mapping is actually a great way to test the Problem Description. It will identify missing objectives. It will challenge the boundaries of the chosen scope (whether too large or too small). It will raise questions that cause you to do more Looking and Asking. So, while it is important to get the Problem Description right to avoid wasting time solving the wrong problem, it is also important not to stall trying to perfect the Problem Description. Instead, feel free to move into Causal Mapping, given it is a lightweight process; just be quick to revise the Problem Description as the Causal Mapping helps reveal where improvement is needed.

Example Problem Description

This is an intentionally oversimplified example (so, keyboard experts, please do not get hung up on the fact that a keyboard would not likely be constructed this way). We simply wanted to craft an example that could easily be analogous to many different things that might exist in your own design space. It is inspired by several real-world problems for which we've coached teams on causal analysis.

Imagine a team responsible for a brand new keyboard design. After successful automated machine testing for accuracy and endurance, the team conducted extensive human user testing and found the problem described in Figure 4.5.

Does that sound like a good Problem Description to you? Is it clear what analysis needs to be done next?

Problem Description

In the development of our new keyboard, we chose to go with single threshold sensors for detecting key presses. As a result, we are now experiencing a 3-6% rate of dropped keystrokes in some of our user testing. Fortunately, our design is compatible with the prior keyboard's dual threshold sensors, so the swap should not be too difficult.

We need this problem solved prior to the end of the month, including re-testing of the keyboard with 0% dropped keystrokes, in order to avoid delaying the release of the keyboard.

Figure 4.5   Example keyboard Problem Description.

Jumping to Solutions

A common failure mode throughout Problem-Solving efforts is jumping to solutions or conclusions prior to proper analysis. And very often that happens in the Problem Description itself. As a test, ask, "Is there more than one way to solve this problem?" If there are not multiple ways to solve a given Problem Description, then most likely what you have is a "Lack of this Solution" Description. When that happens, you want to ask why that solution is necessary or desirable in order to work back out to the real problem. Try not to describe what should exist (a solution) but rather what should not exist (the problem) or the desired end result (what the customer wants from some solution).

Knowing that jumping to solutions or conclusions is a failure mode, often you see people push back: "No, we're not talking about solutions yet." We don't recommend such pushback as it is frustrating for the person with a great idea and it will be difficult for them to engage in proper analysis when they think this is all pointless because they already know the solution.

So instead, we recommend that you jump ahead in the Problem K-Brief (details on the Problem K-Brief later) and capture the solution idea in the Alternatives Evaluation Matrix. After all, we certainly do not want to forget a great idea! But more importantly, you can then ask two key questions that will help develop out the Problem Description.

1. What makes this solution so good? The person who volunteered the idea will certainly be able to answer that. And their answer is likely a list of good objective criteria for your Problem Description (though often already there).
2. What makes this solution *not* so good? What might it be bad at? The champion of the idea may not answer so well here, but others often will. And it's the answer to this question that often uncovers objective criteria that might have otherwise been missed for a while.

With that done, the person with the idea can now relax; the idea has been captured. And the identification of things it is bad at justifies doing some further analysis before settling on the idea, even if it does turn out to be the best choice.

And note that solution ideas may be generated during the Causal Mapping process as well, and we suggest the same response: capture it and use it to prompt for missing objective criteria that should appear in both the Problem Description and the Causal Map.

But certainly do not put solutions in the Problem Description.

Improved Example Problem Description

Can the prior example Problem Description, as worded, be solved in more than one way? Not really. It is in fact a solution (use dual-threshold sensors) masquerading as a Problem Description. Getting the team to look at the situation with an open mind, collecting the larger facts of the situation, you might get the Problem Description shown in Figure 4.6.

But we don't want to lose the potential solution that was in the first Problem Description. So, as suggested in the prior section, that has been captured into the Alternative Remedies section of the larger Problem K-Brief, as shown in Figure 4.7.

In doing so, they asked for evaluation criteria. The original Problem Description suggested that this solution would be implementable by the end of the month, avoiding a delayed product release. But there was also pushback from the team that the dual-threshold sensor was avoided for one reason: part costs.

Problem Description

Our new keyboard was passing all of our automated mechanical testing. 0% dropped keystrokes; 0% sticky key rate; no mechanical failures. So, we began some final human testing... roughly 5% of the human testers reported 3-6% dropped keystrokes. What is different about those human testers vs. our automated mechanical testing? And how do we fix the keyboard prior to our scheduled release end of this month?

The objective is to revise the automated testing to detect the dropped keystrokes, and then modify the keyboard design to not exhibit those dropped keystrokes. Both prior to the end of this month, so as to not delay product release. And without adding more than $1 in part costs.

Figure 4.6   Improved keyboard Problem Description.

Alternative Remedies	Row Status and Reason	Time to Market (wk)	Added Part Costs / Reduced Profitability ($)
	⌄	<= 4wk tested	<= $1 <= $8
Use Dual Threshold Sensors from Prior Keyboard			
Use Dual Threshold Sensors, but lower cost			
Use Single Threshold Sensor (existing design... but fix some other way??)		??	??

Figure 4.7   Moving solutions to the Alternative Remedies section.

The profitability goals for the new keyboard would be tough to achieve with the prior keyboard's sensor setup. And thus, they identified an additional key objective (profitability) that then showed up back in the Problem Description.

So, with consensus on that Problem Description, they began the Causal Mapping process, looking for the root cause(s).

What Is a Root Cause?

As noted earlier, most RCA methods don't have a clear answer to what is a root cause, making it difficult for engineers to know when they are "done." In Problem-Solving scenarios where we want to make changes to fix the problem, we assert that the proper definition of *root cause* is

> **A decision that we can directly control that will impact (either directly or indirectly) the problem or the targeted customer interests.**

If we cannot control it, then it is not the root cause that we are looking for! We can't change it, thus we cannot use it to fix the problem.

In some cases, the distinction can be subtle, but it is very important. For example:

- *Environmental conditions*: We cannot change them, per se; we have to design around them. But we can potentially decide what environment we are designing for.
- *Customer interests*: That is the customers' decision – what they want. We have to design the product to satisfy that. But again, we can decide to what degree we will satisfy those interests and what customers we want to satisfy.
- *Component characteristics*: We can't decide the characteristics of aluminum or the characteristics of the off-the-shelf parts we are using. We need to know those characteristics and work around them. But we can potentially decide which parts or which materials to use.

The point here is to get very crisp in our language and clearly separate what are our direct decisions versus what are the decisions we are making indirectly by making other decisions. I am directly deciding what material to use. But in doing so, I am indirectly picking the cost, density, elasticity, modulus, ultimate strength, and so on of the material we are using. And, in turn, those material characteristics may indirectly impact the product characteristics the customer desires.

The root causes that we are looking for are the decisions that we directly control.

So, the first part of the Causal Mapping process is to ask "Why?" for each identified decision, situation, condition, or sub-problem. And each answer should be added as a shape in the Causal Map with a line to the thing it caused. (See the example in Figure 4.8.)

And when answering "Why?" we should be primarily looking for answers along the lines of

- "Because we decided _____."

But it could also be any of the following:

- "Because this condition _____ existed."
- "Because this situation _____ occurred."
- "Because this problem _____ occurred."

If the answers fall into the last three areas, then we definitely need to keep asking "Why?" until we get to the things we decided. Even when we get to something we have decided, we may want to keep asking "Why?" to see if those decisions depended on other decisions that might be easier to change. But it is okay to stop once you reach decisions that we control; the rest of the Causal Mapping process will generally flesh out the rest of the related decisions in any case.

If this sounds like 5-Why Analysis, it is no accident: we want that same simplicity here. However, we are adding three critical ingredients to simple 5-Why Analysis:

1. A clear terminating condition (decisions you can control)
2. A visual map that supports branching, rather than just a linear progression of Whys
3. A path to establishing a Decision Map to guide decision-making regarding the optimal solution to the problem

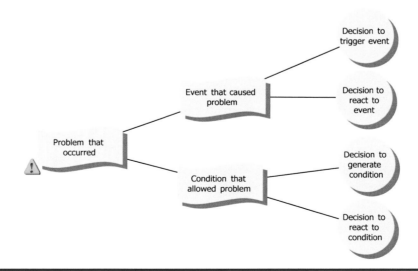

Figure 4.8   Generic RCA as a Causal Map.

To make the distinction between decisions and other things visually clear, we use a circle shape for decisions and a "scroll" shape for all other general causal statements. As we continue to add shapes, the visual tends to grow into a large "tree" – very similar to a Fault Tree. Note that Ishikawa (fishbone) diagrams become awkward if you go more than two levels deep. And yet simple 5-Why Analysis suggests you need to be going at least five deep, typically. Causal Maps can easily go many more than five deep (as you will see in Part II of this book).

Figure 4.8 shows the typical progression from the problem statement (marked with a "!" flag) to the events and/or conditions that led to the problem and then to the various decisions that we made that led to those events or conditions.

Root Causes of the Keyboard Problem

To solve the Keyboard problem, there may be a few different Causal Maps that will be useful to facilitate the LAMDA Discussions and any further analysis that may be needed.

This first Causal Map (Figure 4.9) starts with the problem statement (again, the "!" flag is used to mark problems; the "key" flag is used to mark the key starting point of the Causal Map, so that people know where to start reading). Then,

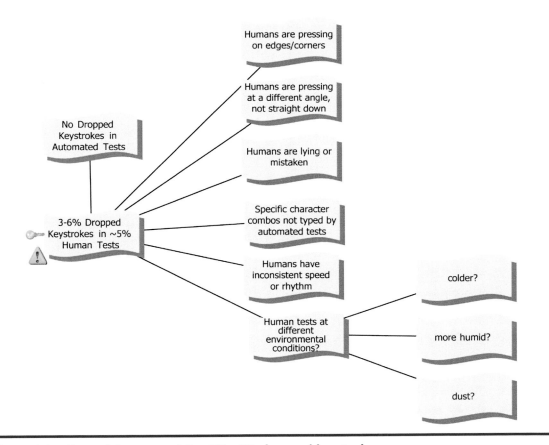

Figure 4.9 Potential causes of the 3%–6% dropped keystrokes.

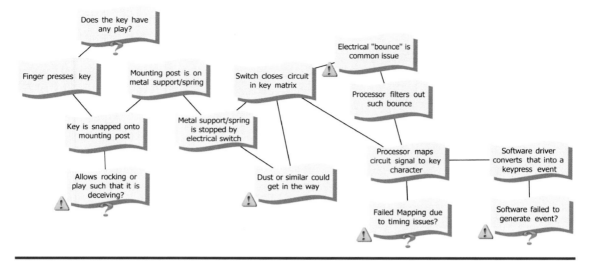

Figure 4.10   Causal Map showing the sequence of causal events.

in the statement above that, it contrasts the problem found in the human testing with the automated results. Based on that contrast, it then brainstorms reasons that the human testing may have differed from the automated testing.

This second Causal Map example (Figure 4.10) lays out the sequence of events starting from the human finger pressing the key. It then brainstorms potential failure modes for each step in that sequence, marking the failure modes with "!" flags. For those familiar with FMEA, this is a lightweight FMEA being leveraged to identify the deeper causal relationships.

The two different Causal Maps represent two complementary lines of exploration:

1. What are the differences between human tests and automated tests?
2. What are the potential failure modes?

Some of the failure modes in the second Causal Map helped us identify potential differences in the tests that were added to the first Causal Map.

At this point, the team decides it is time to stop brainstorming, go *Look*, and do some testing to validate (or deny) the potential causes identified by the Maps. For efficiency, they'll try to make changes to the automated testing to better match human testing and see if they can duplicate the failures. If so, then they can isolate the change that made the difference; if not, then they may need some more brainstorming. Their plans are captured, including "When will Who do What?," as shown in Figure 4.11.

Validating Potential Causes

Another common failure mode of causal analyses is accepting causes as causes simply because it is logical. Just because something certainly *could* cause the problem does not mean it *is* causing the problem in your case. It is critical to find

Done	When ▲	Who	What
☐	*2017-02-21*	Scott	Investigate environmental conditions (go look)
☐	*2017-02-21*	All	Go look at disassembled keyboard (CMap 2)
☐	*2017-02-22*	Jeff	Work up a design for adjusting angle of presses of automated tests
☐	*2017-02-22*	Jane	Re-run automated tests typing same as humans were typing during failures
☐	*2017-02-23*	Jane	Adjust automated tests by misaligning to keyboard
☐	*2017-02-23*	Jerry	Randomize timing of automated tests
☐	*2017-02-24*	Jane	Re-run tests with environment adjusted per Scott's findings.

Figure 4.11   When will Who do What to close the knowledge gaps?

ways to validate that the potential causes are actually causing the problem in your case, otherwise the more critical causes may never get identified. You may fix something that should probably be fixed, but end up delaying the learning that there are yet more causes that also need to be remedied to truly fix the problem.

This is another opportunity to apply the LAMDA enabler – to get the team out of the conference room where they are brainstorming potential causes and in front of the system exhibiting the problem, conducting tests to validate those potential causes.

LAMDA for Causal Mapping

In introducing the all-critical starting point for Causal Mapping, the Problem Description, we provided a table (Table 4.1) suggesting some ways to apply LAMDA to the development of a Problem Description. But LAMDA is applicable to all learning situations – not only the Problem Description but also Causal Analysis, the generation of *Alternative Remedies, Trade-Off Analysis*, the closing of *Knowledge Gaps*, and so on. So, we would like to provide a similar table (Table 4.2) suggesting some ways to apply LAMDA to the Causal Mapping effort.

Validating Causes of the Keyboard Problem

At the next stand-up status meeting of the larger team, our keyboard problem-solving team presented the results of their cause validation efforts, as shown in Figure 4.12.

Based on that testing, it turns out that none of their potential causes seem to be the issue in this case (they are each marked with an "X" to indicate that those potential causes were *disproven*). Fortunately, after the brief presentation of the Problem Description and the Causal Maps to the larger group, Lucy (who was not involved in this keyboard design, but who has worked on prior keyboards) asked if we'd had complaints about the keyboards having poor feel or being tiring to type on. We had seen a few such complaints, but less than

Table 4.2   LAMDA Applied to the Causal Analysis

Look	If questions arise, don't hesitate to go back and look for yourself, with an open mind.
Ask	Ask "Why?" (at least five times). Ask "What actively causes this?" Ask "What conditions must exist?" Ask "When is this more likely to happen?" Ask "Why does this not happen sometimes?" Ask "How can we validate the causes?"
Model	Use a Causal Map to model the science-based cause-and-effects. Use diagrams, charts, and equations to quantify the amount of each effect.
Discuss	Make sure that all understand and agree that all the important causes have been identified and all the effects quantified.
Act	Quickly act on this learning by devising alternative remedies and using these cause-and-effect models to evaluate them.

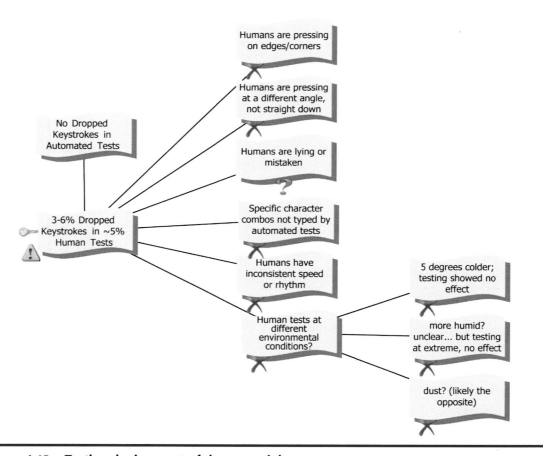

Figure 4.12   Testing denies most of the potential causes.

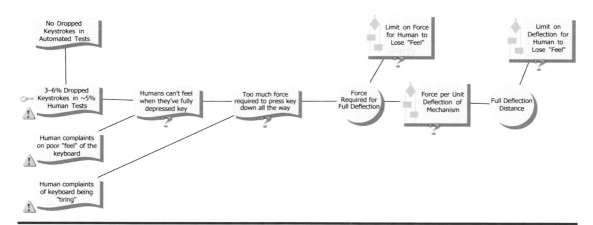

Figure 4.13   Mapping lack of feel to the trade-off between force and deflection.

2%, not enough to warrant immediate action. She said that a few years back they had a keyboard with such complaints and it turned out that if the force required to press the key all the way down was too high, then people would lose the "feel" of whether or not they pressed it all the way down. She suggested that may be the problem here: that they felt they had pressed it all the way down, but had not really. The automated machine doesn't use such feel, it is calibrated for how far to push.

Following that line of thinking toward a root cause, a third Causal Map (Figure 4.13) emerged during that meeting. It has the same problem starting point but adds a couple of additional problem statements (the other keyboard complaints that they had not looked at beforehand). It then identifies the cause as being a lack of feel, which in turn was caused by excessive force being required, which was caused by the measurable design decision of how much force is required for full deflection. Both of those causal statements (the scroll shapes) are marked with a "?" flag since it has not been proven that the problem is due to a lack of such feel; that needs to be validated. Continuing to the right, there will be a relation between that force and the related decision of how much deflection is required for a full press of the key (the greater the deflection, the greater the force required). That relation is not known and so is marked with a "?" flag, indicating it is a knowledge gap. In addition, it is also unknown how big that deflection needs to be to provide good feel. Similarly, it is not known what the upper limit on the force needs to be to provide good feel. Each of those unknown limits are shown as additional limiting relation shapes on each of those decisions, and each relation shape is marked with a "?" flag.

In this Causal Map, you can see the beginning of a Decision Map in the five shapes on the far right. Watch for that to continue to evolve.

Is Knowing the Root Cause Enough?

Most RCA and Problem-Solving methodologies stop the causal analysis at the "root cause" (typically an ill-defined root cause) and then move on to the ideation of potential solutions to the problem. However, knowing all of the "root causes" is *not* enough to know how to fix the problem. That tells us which decisions we should consider changing, but it does not tell us what to change them to.

To know how much we can change any one of those "root causes," we need to know the impact that changing that decision will have on all other customer interests.

So, the next phase of the Causal Mapping process involves asking not "Why?" but rather "So what?" or "What other customer interests will be impacted if we change this decision?"

Each answer should be added to the Causal Map as an appropriate shape until you get back out to a *customer interest decision* (i.e., our decision as to what degree we will satisfy the specified customer interest). So, Causal Maps tend to start with customer interests that are not being satisfied (problems or key objectives), work out to the root causes (the decisions you control), and then continue to work out further to the other customer interests that may be impacted (the trade-offs that need to be considered when fixing the problem or improving the objectives).

It is critical that teams embrace these customer interests on both sides as decisions that they control, rather than as inputs to the process; otherwise, there is a tendency to accept those as known and achievable, when in fact they very often are not achievable, and what part of the achievable space is most desirable to the customer is not really known.

For example, a customer may ask for a jet engine with at least a certain thrust but which weighs less than a certain weight and is less than a certain length. The jet engine company may not know how to deliver that much thrust out of an engine that small. So, should they hit the thrust level by violating the length or by violating the weight? Or should they hit the length and weight (so it fits on the aircraft) and give the most thrust possible (though less than what was asked for)? Either may be completely unacceptable to the customer. The most desirable design for the customer may be to compromise all three by some amount. That may be very difficult for the customer to decide without guidance on what is actually possible. Said another way: until the customer is equally informed, the customer and the jet engine supplier will likely continue to disagree (the customer demanding the impossible and the jet engine supplier rejecting the requirements as impossible)! Hence the need for Trade-Off Charts (the topic of the next section): to get everyone equally informed.

Failure to identify such effects of changing a decision (a root cause) can result in not identifying key decisions and the associated knowledge gaps until very late in the process, and thus causing some of the worst loopbacks (the most expensive rework).

To make the critical distinction between customer interest decisions and normal design decisions visually clear, see Figure 4.14, where circle shapes for the customer interest decisions are tagged with a red target symbol.

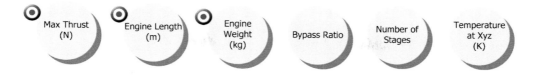

Figure 4.14 Decisions are shown with circular shapes (targets mark customer interests).

So, as the Causal Map continues to grow, it becomes less a traditional tree structure and more a collection of interconnected shapes surrounded on all sides by customer interest decisions (since that is the terminating condition for this second phase of the Causal Mapping process; more on that will follow).

Is Knowing All the Impacts Enough?

When considering a change to a decision (a root cause) in order to fix a problem, it is certainly helpful to know the other customer interests that you will impact. But unless you know to what degree you will impact them, you still do not know how much you can or should change each decision. To know that, you need to know the sensitivities.

As I change this decision, how much impact does it have on each of the customer interests?

Without knowing that, you do not have visibility to the trade-offs you are making on behalf of the customer. To reduce the maintenance costs by this much (the problem I am trying to fix), how much am I affecting the *X* performance, the *Y* performance, the weight, the purchase price, and so on?

Not knowing those trade-offs not only prevents you from making the decisions you need to make, it also prevents you from even learning where the customer would prefer to be in that multi-dimensional design space. You can't ask customers to prioritize every possible point in that space; that would be beyond overwhelming and impossibly impractical. You have to show them the specific trade-off curves that exist and let them answer where they would like to be on those curves. Even having that much easier question answered can be challenging enough.

If you ask engineers for the trade-off curves between the customer interests, they can generally give you a rough notion. They know if the curve goes up or down (or both) and whether it's fairly linear or has a sharp knee. But if you ask for more precise numbers, they will answer that it depends on a lot of other things. To which we respond with a question: "What other things does it depend on?"

We add those answers to the visual model, the Causal Map, as additional decision shapes or causal shapes hanging off the Relation shapes that are dependent on those decisions.

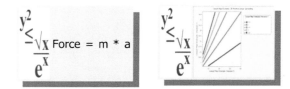

Figure 4.15   Relations between decisions are shown with rectangle shapes.

Once you have those dependencies in the Causal Map, you can start to find out what the trade-off curves are between small subsets of those shapes. By breaking things down in this causal analysis process, you eventually get down to relations that the engineers know. Many will be very simple.

- This is just the sum of these other three (Costs, weights, lengths, etc.).
- This is just the product of these other two ($F = m \times a$; $F =$ spring constant $\times$ spring displacement; Weight = density $\times$ volume; etc.).
- This is less than or equal to that (Energy Out $\leq$ Energy In; weight $\leq$ lift; drag $\leq$ thrust).

Others may be complex physics equations. And still others may not have knowable equations; they must be tested experimentally.

- We would need to determine that with wind tunnel testing.
- We can do a CFD or FEM analysis to determine that.
- We have a 6DOF Simulation that can tell us that.

And in still other cases we may have nothing more than rules of thumb or historical trends to go by. Where did our past jet engine designs end up?

For each such relation that we know among a subset of the decisions, we make that visible in the Causal Map via a rectangular relation shape, as shown in Figure 4.15. (We use rectangles because we often draw the Trade-Off Chart inside the rectangle, and charts are naturally rectangular.)

Continuing the Causal Map of the Keyboard Problem

To make this discussion a little more concrete, let's continue the Causal Mapping effort for our Keyboard Problem, covering the latter two phases of the analysis, from root causes back out to the other customer interests and the trade-offs between them.

When they asked themselves about the trade-off between *Force* and *Deflection*, they responded that the trade-off depends on the geometry of the spring mechanism and the material properties of the spring mechanism. The geometry of the spring mechanism is not a single measurement with a specific unit of measure, it is multiple different design decisions. Similarly, there are multiple material properties that will need to be known. In Figure 4.16, we use a cloud shape (i.e., a *fuzzy statement*) to indicate "multiple of the other three

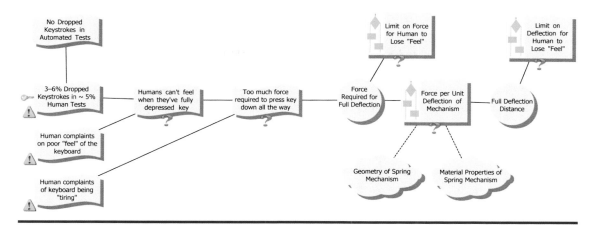

Figure 4.16   Zooming in on the causal decisions and the relations between them.

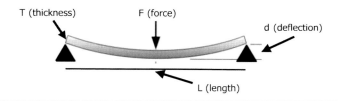

Figure 4.17   Rough visual model of the spring mechanism.

shapes go here" (circular decisions, rectangular relations, or scroll-shaped general causal statements). It could be that we plan to fill them in later. Or it could be that those details are given in a separate Causal Map.

To break out those two fuzzy statements, they decided a visual model of the spring mechanism would be helpful (Figure 4.17).

Zooming in on just that portion of the Causal Map, they detailed out the two fuzzy statements as shown in Figure 4.18.

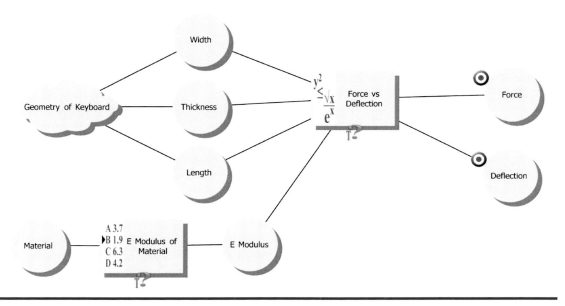

Figure 4.18   Expanding the Causal Map to the decisions we control.

They felt they could find an equation for that *Force-versus-Deflection* relation for that simplified model (a simple beam), though they would need to do some testing to verify that it applied to the actual part that was a little different. And they knew they could get the E-Modulus for the material used and for several of the known alternative materials. So they marked each of those relation shapes (rectangles) with a "T?" symbol, indicating "We think we know this, but we need to go find that knowledge" (more on that notation in the next section).

Identifying the Knowledge Gaps

In addition to making clear what we need to know to change a decision (a root cause) to fix the problem, the Causal Map also makes visible what we do not yet know.

- I know there is some equation that relates these four things, but I don't know what it is.
- I know we can test the relationship between these five things, but we haven't yet done that testing.
- I know these three things will affect this, but I'm not sure on this fourth, and there may be others.
- I know this will be impacted, but I don't yet know how we would measure that (e.g., ease of use, ease of maintenance, visual appeal).
- I think this might be a causal condition, but I am not sure; we should validate that.

To make such knowledge gaps visual, we mark the shapes that have associated knowledge gaps with a question mark (?). Further, it is often useful to distinguish the type of knowledge gap by putting one of three letters next to the question mark (as shown in Figure 4.19):

- T? = *Think We Know*: We just need to go find the right person or the right data.
- O? = *Opinion*: This is someone's expert opinion, but they have no data to back it up; further analysis or testing may be desirable.
- N? = *Need to Know*: Neither of the above, we just need to go learn this.

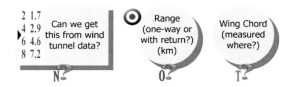

Figure 4.19   Flagging knowledge gaps with N, O, or T (maps are KNOT Charts).

Adding this distinction is not necessary to the process, so an unadorned question mark is also acceptable, which just implies you are not classifying the knowledge gaps that way. (With the knowledge gaps marked this way, the Causal Map is effectively a *KNOT Chart*, where the things we *Know* (K) are shown with no question mark at all.)

Finding the Unknown Unknowns

When introducing the power of the Causal Map to identify knowledge gaps to organizations, we are often asked, "That's great, but what about all the knowledge gaps that I don't yet have in my Causal Map because I don't know to put them there? What about the *unknown unknowns*? Those are usually the most costly."

First, that's why it is important that the Causal Map is a very simple visual model that can be presented quickly to anybody without training because we want to get as many possible experts with as many possible experiences to look at the Causal Map, understand all the interrelationships that have been identified so far, understand all the customer interests that may be impacted that have been identified so far, and then brainstorm what's missing. The more experts you have efficiently critiquing the Causal Map, the more likely you are to find what's missing.

Second, the Causal Map is actually designed to make visible when things are missing. For example, one of the steps in the Causal Mapping process (see Appendices III through VI) involves asking the question "Are there dangling decisions?" That is, are there decisions that impact just one customer interest rather than being in between two or more? If so, then we can generally tell you that there's something missing from your Causal Map without even knowing anything about your design space.

How do we know that? Because solving this problem would be really easy otherwise. If you are trying to reduce Cost, and there's a decision (*Length* in Figure 4.20) that causes Cost to go down as it goes down, then you can simply drive it to near-zero to drive the Cost to near-zero. Easy!

Of course, the response is usually, "You can't make that really small (or huge)!" And we ask "Why not?" And then they tell us a bunch of things that will be impacted that are not yet in the Causal Map. For example, "The Length is providing the leverage that allows the motor to lift the object." So, missing from the

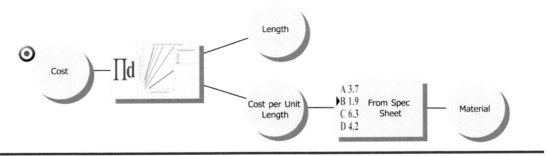

Figure 4.20　Length is a dangling decision.

preceding Causal Map is the customer interest decision of how much weight can be lifted, the computation of the leverage effect, and the trade-off with the motor size, which probably also affects Cost.

In that way, the visual model helps you to identify the unknown unknowns!

But with that said, it is certainly still possible to miss some, resulting in rework. When that happens, we want to make sure we can continuously improve our knowledge so that failing to identify the issue will only ever happen once, and from then on that learning will be baked into our decision-making.

Incorporating these Decision/Causal Maps into your organization's development system best practices are key to ensuring that you will never miss a specific unknown unknown more than once. Experience has shown that while developing the maps, the teams will become better at identifying the unknown unknowns in advance of them becoming rework.

Unknown Unknowns in the Keyboard Problem

Are there dangling decisions in the last Keyboard Map? Yes: the Material decision is not a customer interest, and it is bounded on just one side by a customer interest. So, to improve the Force-versus-Deflection trade-off, they could just pick a material with a super-low E-Modulus. So, why couldn't they do that?

The answer is that such materials would likely not have the required durability. If they want the keyboard to last forever, then the Stress caused by the key presses must never exceed the endurance limit of the chosen Material. They added that to the Causal Map (the blue flags in Figure 4.21 indicate the shapes added).

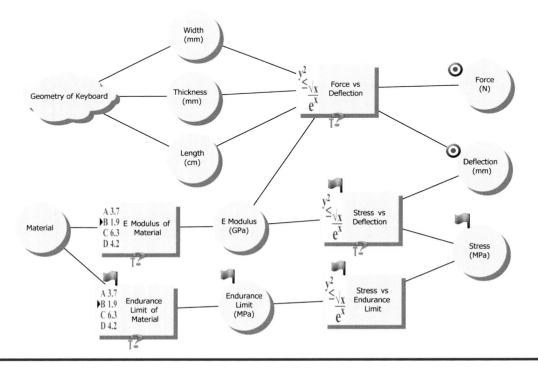

Figure 4.21   Adding Stress and the Endurance Limit to the map.

Again, they thought they could go find an equation for that Stress for a rectangular beam, and they knew they could find the Endurance Limits of the materials (and the known alternatives). With that, they have completed the second phase of the causal analysis: they have identified all the customer interests that may be impacted by changing the material selection decision (a root cause). To complete the causal analysis, they need to actually find those equations and material properties so that they can compute the Trade-Off Charts (to compute the Force-versus-Deflection trade-off while never exceeding the Stress-versus-Endurance limit).

Closing K-Gaps for the Keyboard Problem

The equation for Force-versus-Reflection for a rectangular beam is shown in Equation 4.1.

$$F = \frac{4E \times W \times T^3}{L^3} d \tag{4.1}$$

where:
   F is the Force
   d is the Reflection
   E is the E-Modulus
   W is the Width
   T is the Thickness
   L is the Length
   (as was depicted graphically in the visual model of the physical spring mechanism shown earlier)
   The equation for Stress-versus-Deflection is shown in Equation 4.2.

$$S = \frac{6E \times T}{L^2} d \tag{4.2}$$

Based on this, they computed the Trade-Off Chart of Force-versus-Deflection shown in Figure 4.22.

The current material is the Tempered 1 in green. A known workable alternative is Alloy 1, and notice it has a much lower Force for given Deflection than the rest. Perhaps that would be adequate to give the desired "Feel"; only human testing would say for sure. Note that Alloy 1 cuts off at a smaller Deflection than the current Tempered 1 material; but as long as the required Deflection is less than that cutoff, Alloy 1 should work. Why is it cut off? A Trade-Off Chart of Stress-versus-Deflection will make that clearer (Figure 4.23).

Although the Alloy 1 Stress-versus-Deflection slope (purple line) is lower, it runs into a much lower Endurance Limit cutoff (purple shaded region), where the Stress would exceed the Endurance Limit. Hence, the purple line stops at a lower Deflection than the green line in the previous chart.

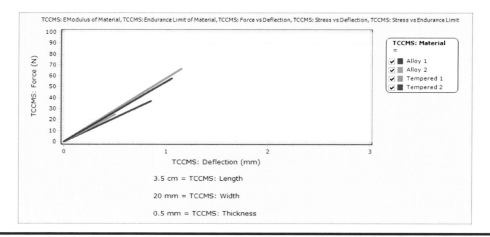

Figure 4.22 Trade-Off Chart of Force vs. Deflection for different materials.

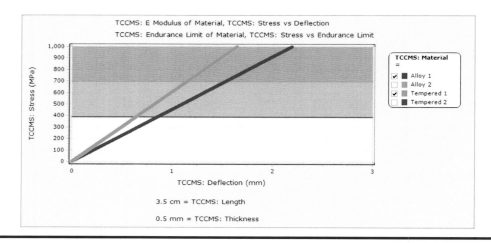

Figure 4.23 Trade-Off Chart of Stress vs. Deflection and the Endurance Limit.

These visual models confirm that a feasible design exists that will improve "Feel"; human testing is still needed to determine how low a force and how large a deflection is needed to give adequate Feel. Given the timeline, a quick mock-up might be produced with the improved design, with the hope that it would pass human testing. But with a little more time, it would be better to do Set-Based testing to fully characterize the customer interest in feel once and for all, such that future designs can be better optimized. How low does the Force need to be to provide adequate Feel?

If the team's schedule did not constrain them to the pre-approved list of materials, then they could go out looking for materials with a lower E-Modulus and a higher Endurance Limit. But when they do so, they may start finding higher cost materials or materials that are more difficult to work with in manufacturing. To evaluate such alternatives, an expanded Causal Map would be needed (the six shapes on the left side of Figure 4.24 marked with blue flags were added).

That expanded Causal Map may not be needed for the current decisions to be made. But note that once the expanded Map is developed, it remains useful

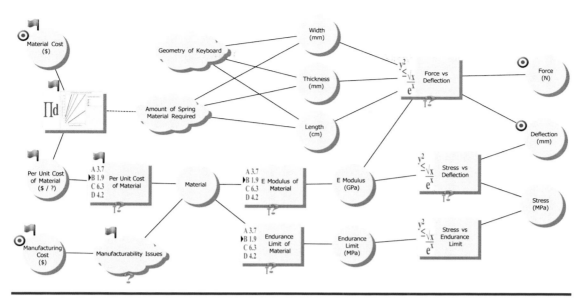

Figure 4.24 Mapping further to Material Cost and Manufacturing Cost decisions.

and applicable every time we need to make decisions in this space. So, even if it is not needed now, capturing this expanded Map (as it arises in the LAMDA Discussions) is valuable, as it provides a better starting point for all future work. The next project may not need all of this either, but perhaps one of those clouds may be the key reminder they need to not get bitten by an unknown (to them) unknown.

Causal Map versus Decision Map for the Keyboard Problem

Although we hope we have made the distinction between the Causal Map and Decision Map clear, experience has shown us that it helps to repeat the distinction now that we can provide a concrete visual model, as shown in Figure 4.25.

The Causal Map is a simple visual model designed to facilitate collaboration among people with different expertise. By sticking to just four simple shapes (circles for decisions, rectangles for relations between decisions, scroll shapes for general statements, and clouds to represent where multiple other shapes belong) and a few flags on those shapes (targets for customer interests, question marks for knowledge gaps, keys and other flags for highlighting), the Causal Map remains simple to understand, even for those with no training in it.

For those trained in it, they are searching for the key decisions that need to be made and the impact that each decision will have on the others (in particular, the customer interest decisions). And thus, subsets of the Causal Map will emerge where there are only decision and relation shapes. Such a subset we refer to as a Decision Map since it can be used and continuously improved as a team's best practices for making that set of decisions.

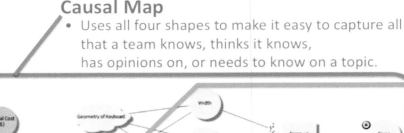

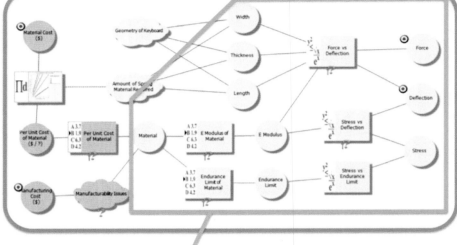

Figure 4.25   Distinction between a Causal Map and a Decision Map.

Visual Models beyond Causal and Decision Maps

Although the focus of this chapter has been the new visual models, the Causal Map and the Decision Map, we don't want to leave you with the impression that they are the only visual models you need to add to your LAMDA Discussions. In fact, there are many visual models that are commonly needed.

■ Annotated Pictures, Diagrams, and/or Timelines to better communicate the Problem Description or topic of Concern
■ Ideation tools like Nine Windows to help identify all the potential Contributors to the Problem or topic of Concern
■ Pareto Charts to identify the most prevalent contributors (and thus those that should be included in the Causal Map)
■ Alternatives Matrices to identify the solutions that are generated, but more importantly, to help identify the objective criteria that you will use to evaluate those alternatives (and thus should be included in the Causal Map)
■ Free Body Diagrams, Kinematic Diagrams, Wiring Diagrams, Circuit Diagrams, Process Flow Diagrams, and any other such diagrams that show the important physical elements of the system, how they physically relate, and how they are measured (typically giving clarity to the Decisions identified in the Causal Map)

That larger set of visual models should all be visible to those participating in the LAMDA Discussions developing the Causal Map. That is the role of the K-Brief enabler described in Chapter 3. It should organize those visual models in order to most clearly tell the story that you want all participants in the LAMDA Discussion to understand. The visual models are key to making sure all participants have a shared understanding, which is key to efficient and effective Discussion. The Causal Map might be the most valuable of those as it is the most rigorous, but other visual models are often needed to support the Causal Map. For example, consider the Causal Map from the example Keyboard Problem shown in Figure 4.26.

Without the simple annotated drawing shown in Figure 4.27, the Causal Map in Figure 4.26 would be a lot less meaningful to the LAMDA Discussion participants.

Further, the Problem Description and Trade-Off Charts found in the larger Problem K-Brief explain the significance of the Force and Deflection decisions. All together they facilitate efficient LAMDA Discussion and consensus decision-making: *Equally informed people seldom disagree.* The K-Brief coordinates the use of visual models to ensure everybody involved in the LAMDA Discussion is equally informed.

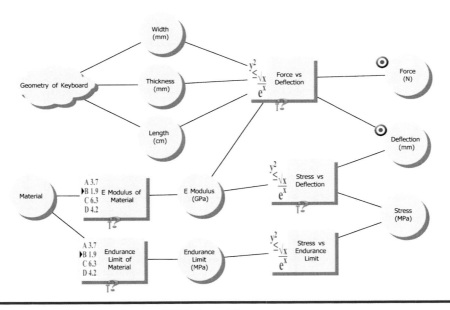

Figure 4.26   Causal Map for the keyboard problem.

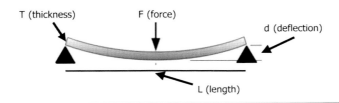

Figure 4.27   Complementary visual model of the spring mechanism.

PROBLEM

TCCMS: 7 Keyboards are passing automated testing but human testers complain they drop keystrokes

LEAD	ID
TCC Admin	1049960
LAST EDITED ON	STATUS
2017-03-24 13:03	Draft
CATEGORIES	
Material Selection Example (TCCMS)	

ABSTRACT

This is an example Problem... it is intended to show Causal Analysis brainstorming potential causes... testing confirming or denying those... LAMDA Discussion helping to brainstorm more... and ultimately digging to root causes (decisions we control).

Problem Description

Our new keyboard was passing all of our automated mechanical testing. 0% dropped keystrokes; 0% sticky key rate; no mechanical failures. So, we began some final human testing... roughly 5% of the human testers reported 3-6% dropped keystrokes. What is different about those human testers vs. our automated mechanical testing? And how do we fix the keyboard prior to our scheduled release end of this month?

The objective is to revise the automated testing to detect the dropped keystrokes, and then modify the keyboard design to not exhibit those dropped keystrokes. Both prior to the end of this month, so as to not delay product release. And without adding more than $1 in part costs.

Causal Analysis

It may be useful to look at potential differences between the human testing and the automated testing.

It may be useful to look at the sequence of events and brainstorm failure modes. (Some of the failure modes here helped us identify potential differences in the tests listed above.)

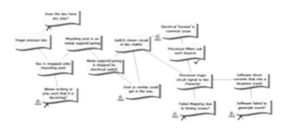

In the weekly cross-team review, Lucy asked if we've seen complaints on the keyboards having poor feel or being tiring to type on. We had seen 2% of such complaints... not enough to warrant immediate action. A few years back, they had a keyboard with such complaints and it turned out that if the force required to press the key all the way down was too high, then people would lose the "feel" of whether or not they pressed the key all the way down. She suggested that such may be our problem here... they felt they pressed it all the way down, but hadn't really.

So, following that line of thinking towards root cause, this initial Causal Map emerged. We plan to continue the Causal Mapping effort from here.

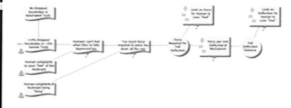

A Simplified Model

For this initial design study, a simplified model of the system was chosen (a rectangular beam supported on both ends with force applied to single point in center) to explore the trade-offs between the material selected and the other design decisions, including:

- d = Deflection of the material due to the force (in)
- F = Force applied to the center of the beam (lbf)
- L = Length of the beam (in)
- W = Width of the beam (in)
- T = Thickness of the beam (in)

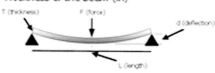

Figure 4.28 Problem K-Brief tells the problem-solving story with visual models (first half).

7) What else might these things depend upon? What other customer interests might depend upon these things?

Notice the Material is dangling... we can solve this Problem by picking some material with super low E Modulus... so, why can't we?

Changing the spring material to get more Deflection for less Force may result in a more flimsy material that wears out more quickly. Ideally, the Stress due to Deflection should never exceed the Endurance Limit of the material.

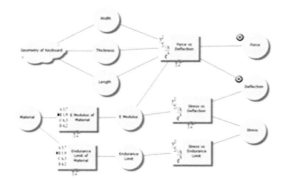

8) Construct the Decision and Relation K-Briefs (construct a Decision Map) so that you can view Trade-Off Charts

Scott was able to collect data on the spring material used, and some alternatives. He also found the following two equations which Jane was able to verify accurately model the existing part through physical testing.

Force vs Deflection

This formula was pulled from XYZ by Joe Johnson:

$$F = \frac{4\,E \cdot W \cdot T^3}{L^3}\,d$$

Maximum Stress vs Displacement

This formula was pulled off ABC textbook:

$$S = \frac{6\,E \cdot T}{L^2}\,d$$

Trade-Off Analysis: Force vs Deflection

Alloy 1 offers a better Force vs. Deflection curve than the rest; but it cannot withstand as large of Deflection as the Tempered Steel 1 we are currently using.

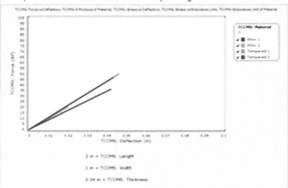

Trade-Off Analysis: Stress vs Deflection

Rearranging the Chart to show the Stress vs. Deflection allows you to see why... the much lower Endurance Limit of Alloy 1. However, as long as we don't need that much Deflection for adequate "Feel" of travel, then Alloy 1 would be the preferred choice.

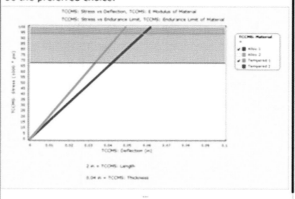

Plan to Close Knowledge Gaps

* To Do once K-Brief's Status >= Discussion

Done	When	-	Who	What
☑	2017-02-21		Scott	Investigate environmental conditions (go look)
☑	2017-		All	Go look at disassembled keyboard (CMap 2)
☑	2017-02-22		Jeff	work-up a design for adjusting angle of process of automated tests
☑	2017-02-22		Jane	Re-run automated tests typing same as humans were typing during failures
☑	2017-02-23		Jane	Adjust automated tests by measuring to keyboard
☑	2017-		Jerry	Randomize timing of automated tests
☑	2017-02-24		Jane	Re-run tests with environment adjusted per Scott's findings.
☑	2017-02-28		All	Continue Force vs. Deflection vs. "Feel" Causal Mapping
☑	2017-		Scott	Got the equations and material data.
☑	2017-		Jane	Test the material against the equations.
☐	2017-03-10		Jane	Test a few keyboards modified with Alloy 1 spring.
☐	2017-		Scott	Setup human testing of the modified keyboards to test "Feel" issues.

Trade-Off Analysis: Titanium? Others?

Titanium offers a much better Force vs. Deflection curve than the rest, and its Endurance Limit is still high enough that it offers better Deflection than most. So, if Alloy 1 "Feel" turns out to be inadequate per the testing, we could consider Titanium (or other more exotic materials).

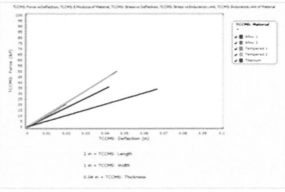

Figure 4.29 Problem K-Brief tells the problem-solving story with visual models (second half).

The Problem K-Brief: Organizing the Visual Models

We started the Causal Mapping process with the Problem Description. Although, logically, it is part of the larger Problem-Solving process and an input to the Causal Mapping, the reality is that developing the Problem Description is part of what a good Causal Analysis process needs to be able to do.

Similarly, throughout this chapter we have shown a number of visual models that are needed to properly support the Causal Mapping process, even though, logically, those visual models are really part of the larger Problem-Solving process. For example, the Alternatives Evaluation Matrix, the physical model of the spring mechanism, and the Trade-Off Charts.

Organizing those various visual models for efficient presentation to the various experts that you want to engage in the ongoing (LAMDA) Discussion is an important part of the efficiency of the process. The K-Brief is the enabling tool for that. To make that a little more concrete, Figures 4.28 and 4.29 show the left and right sides of the Problem K-Brief that would have been evolving through the Keyboard Problem-Solving effort illustrated in this chapter.

Broader Understanding: The Big Picture

The prior section should have given you a much deeper understanding of the Causal Mapping methodology and the details of how to construct a Causal Map, evolve it into a Decision Map, and use it to solve a Problem within the context of a Problem K-Brief. But remember the role of this Causal Mapping methodology: it is the enabler of most of the enablers of "True North" and "Success is Assured" decision-making.

This section will give you a broader understanding of how Causal Mapping impacts the larger development process and literally serves as the "Big Picture" for that process.

Causal Mapping Is Not Just for Problem-Solving!

We have taught our Causal Mapping methodology in this chapter using a problem-solving example for simplicity. However, we want to stress that Causal Mapping is not just for problem-solving. Let us do a quick review of some of the key concepts discussed in this chapter, but in the larger context of product development (or even general decision-making).

First, **the "Causal" in Causal Mapping should not be taken as the "cause of a problem"; rather, it is referring to the causal relationships between the design decisions you are making and the customer interests you are trying to satisfy** (or even more generally, the cause-and-effect relationships between whatever decisions you are making and whatever objectives you are trying to achieve by making those decisions). The only way we can properly make decisions is if we understand the effects that making those decisions will cause. That is what Causal Mapping is about.

Our terminating condition for the first third of the causal analysis is when you get to the root cause(s), which we defined as a decision that you control. That definition works just as well for general product development. If you are not "solving a problem" but rather trying to design a product that will generate the maximum amount of profit, then you need to work back from the profit objective to the decisions that you control that will cause that profitability.

But you cannot stop there: the second third of the causal analysis is to ask what else might be impacted by changing those decisions that affect profitability. There may be other things you care about that might be made worse as you try to increase profitability. For example, those decisions might cause market share to go down.

Once those other impacts are identified, you will want to understand the trade-offs between them in order to make the right decisions. Thus, the three thirds that we described in the context of problem-solving apply equally well for any complex multi-objective decision-making.

The Causal Maps, Decision Maps, Trade-Off Charts, and other visual models were all collected together into a Problem K-Brief in our example. The Problem K-Brief is just one type of K-Brief — a K-Brief for telling a problem-solving story — for pulling the right thinking and collaboration for a problem-solving effort. For a project to develop a more profitable jet engine, you would use a series of Integrating Review K-Briefs to pull the right thinking and collaboration leading to each Integrating Event to make the decisions laid out by the Project K-Brief, which itself pulled the right thinking and collaboration to layout that sequence of decisions. The visual models establishing that "Success is Assured" for each set of decisions would be organized into those K-Briefs.

But that leads to an important complementary point: done right, problem-solving in product development *is* product design!

Problem-Solving Is *Product Design!*

If we are having a problem with our design, it is because we made an incorrect design decision earlier in the development process. To fix the problem, we need to change that decision. But in doing so, we need to keep in mind all the knowledge and trade-offs that went into making the original decision. If we ignore that knowledge and just change the decision, then we are at risk of violating the original design intent, violating key assumptions, or otherwise resulting in causing some other problem when we fix this problem.

We are asking those problem-solvers to re-make the same set of decisions – but to do it better this time and in the face of more constraints (you're too late in the process to change many things that you could have earlier)! Without the Decision Map that captures the logic that went into those decisions, we are really asking those problem-solvers for the impossible. The only reason we don't fail at this more often than we do is that we actively try to change as little as possible (suppressing potential innovation), so we tend to not violate too much of the original design. But without actually knowing the impact our changes will have on the trade-offs, we are essentially guessing and hoping to get lucky. And as things get more complex and more constrained, we will get lucky much less often.

Efficiency: Getting Started and Knowing When We're Done

In working with many engineers in many organizations across many industries, we have found two key failure modes: getting started and knowing when we're done.

For efficient Causal Analysis, you need to be able to answer these four questions:

1. What is a root cause? (When have I gone deep enough?)
2. Do I have all the root causes? (When have I gone broad enough?)
3. Which root cause should I change? (When do I know enough?)
4. What visual model should I be putting on paper to help my team answer any of these prior questions?

If you can't answer the questions in parentheses, then you can't answer "When am I done?" If you can't answer the last of these four, then you can't answer "Where do I start?"

Hopefully, this chapter has given you clarity on all of that. But until you have applied it to some real-world problems in your own space, the power of that to elevate and accelerate your collaborative learning, analysis, synthesis, and problem-solving will remain somewhat unclear. In other words: go do it! (More on that in Chapter 6.)

Efficiency: Decision Maps Are Reusable Knowledge

For example, the Decision Map subset of the Keyboard Problem corresponding to the Force-versus-Deflection Trade-Off Chart is shown in Figure 4.30.

This Decision Map can be used to drive your decision-making in an ongoing way. You will be able to reuse portions of that Decision Map in other Causal Mapping situations, to help drive other decisions that overlap with these decisions. We are not capturing knowledge for knowledge's sake; we are capturing the knowledge that we want to make sure is reused whenever we are making decisions.

Thus, you will tend to grow and improve that Decision Map to cover additional problem analyses or to cover different parts of the design space. In other words, the Decision Map tends to be continuously improved over time as your best practice for making that set of decisions. We are capturing our decision-making best practices, not just general knowledge.

The Decision Map can also serve as guidance for your Research and Development and Technology Development efforts: it gives them clarity on what

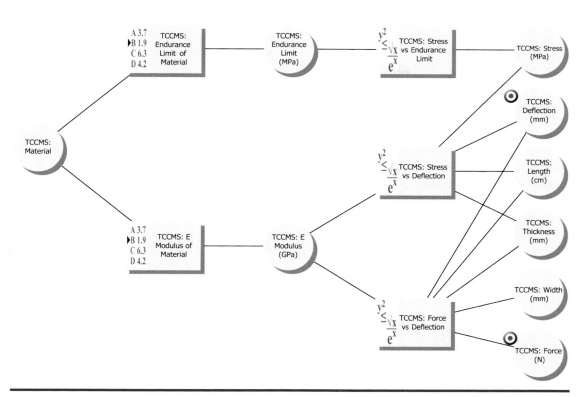

Figure 4.30 Decision Map for the keyboard problem.

knowledge they need to provide for new technologies or new approaches such that those innovations will impact your product teams' decision-making.

The Decision Map can serve as your primary source of *Knowledge Reuse*, serving the same role that Toyota's Engineering Checksheets serve for the Toyota Product Development System. In fact, that was the inspiration for our Decision Map, which in turn was the inspiration for the Causal Map, as discussed at the beginning of this chapter.

Three Layers of Reusable Knowledge

Given the use of the preceding Problem K-Brief to successfully solve the Keyboard Problem, you have effectively built three layers of reusable knowledge:

1. The Decision Map, including the set of Decisions that need to be made with their units of measure and the Relations between those Decisions, which represent the limits on those Decisions and the trade-offs between those Decisions
2. The Trade-Off Charts built from the Decision Map that were the most useful in making the specific decisions for the Keyboard design and will tend to be most useful on future keyboard designs
3. The K-Brief with the larger logic flow and larger set of visual models that together guide a team through the decision-making process based on those Trade-Off Charts, in turn based on that Decision Map

Said another way:

1. The Decision Map makes visible the *structure* of the design space, including *what* decisions need to be made and *what* knowledge needs to be considered.
2. The Trade-Off Charts make visible the *limits* of the design space considering all that knowledge, including the *sensitivities* between the decisions.
3. The K-Brief makes visible *how* the team leveraged that generalized knowledge by using the Trade-Off Charts to converge to the decisions for their specific situation. That logical flow from the general design space to a specific decision is typically reusable no matter where in that design space you need to be.

Together, those three layers of reusable knowledge shown in Figure 4.31 form the Keyboard Team's best practices for making the decisions for which their team is responsible, in the form of tools that they can use to make those decisions. In other words, it is not just reading material for the team; rather, it is the actual tools that they can duplicate and reuse to make the key trade-off decisions on each future keyboard.

Contrast that to traditional knowledge management approaches that either involve capturing best practices as a series of steps or involve capturing lessons learned.

Where best practices are captured in engineering organizations, they tend to be specified as series of steps: do these steps and you will end up with a good design. And then those assigned to review the designs against those best practices tend only to verify that the proper steps were performed, with the expectation that the answer must be right if they followed the steps faithfully. With a Decision Map, you are not checking the steps, you are checking that the decisions themselves do not violate the known limits, represented explicitly in the Decision Map. Further, with the Trade-Off Charts, you are seeing where the chosen solution is in the larger design space. The net result is a much richer design review that has the reviewing engineers thinking about where you want to be in the design space and what limits might be violated. That can lead to not only the identification of mistakes but also the identification of opportunities for further optimization and innovation. For most engineers, that is far more enjoyable than reviewing process steps, meaning they are far more engaged and thus far more likely to contribute positively.

Lessons learned are typically captured in standardized textual documents that can be searched via keywords and part numbers and such. When starting a new design, engineers are expected to use the resulting database to ensure they are incorporating all of those lessons learned into their designs and thereby avoid

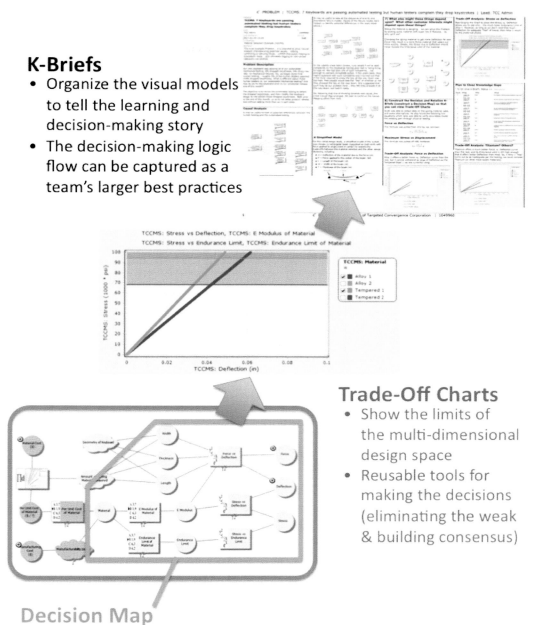

K-Briefs
- Organize the visual models to tell the learning and decision-making story
- The decision-making logic flow can be captured as a team's larger best practices

Trade-Off Charts
- Show the limits of the multi-dimensional design space
- Reusable tools for making the decisions (eliminating the weak & building consensus)

Decision Map
- Captures the structure of the design space (trade-off space)
- Forms a team's best practices for making that set of decisions

Figure 4.31 Three layers of reusable Set-Based knowledge.

making the same mistakes. Unfortunately, that process has several unavoidable failure modes:

- It is easy to choose the wrong keywords and not find the knowledge you need.
- Broader keywords may find far more knowledge than the engineers can wade through to find what is really relevant.

- Determining relevance can often be difficult, as the knowledge tends to be point based and very specific to the situation or parts that they were developing.
- Often, the knowledge found from different parts or situations seems to conflict and the underlying root causes for those differences is not clear (since the prior knowledge capture effort was just capturing what they learned in their specific situation).

Any one of those failure modes can prevent effective reuse. And typically, most of those will be getting in the way all at once. In contrast, by connecting Set-Based knowledge to the decisions that you make on every design, there is no need to choose the right keywords to search for; teams know the decisions they are responsible for making and have fairly consistent and stable naming for those decisions. Further, the Set-Based nature of Trade-Off Charts makes the knowledge applicable to all cases, not just individual points. And finally, the Decision Map captures the full connectivity such that the relevance is always clear.

That is actually a key part of the original motivation for the design of the Decision Map and the broader methodology taught in this book. It is no accident that this book ties everything back to the decisions that your team is responsible for making. Those decisions form a relatively stable backbone for ongoing collaboration, continuous improvement, and knowledge reuse. From project to project, the points where you end up in the design space may be dramatically different, the limits you are running into may be different, and even some of the decisions that need to be considered may change, but the vast majority of decisions that will need to be made will be common. Changing technologies may have dramatically different trade-off curves, but most of the trade-offs will be there in one form or another.

That Reusable Knowledge Is Needed for **This** *Project*

Earlier, we noted that we are asking problem-solvers to re-make the same set of decisions as the prior design team, but to do it better this time and do it in the face of more constraints. Without those three layers of reusable knowledge that capture the logic that went into those decisions, we are really asking those problem-solvers to redo all the prior work from scratch.

In other words, those three layers of reusable knowledge that our projects are building are not just for making the *next* project go more smoothly; those three layers of reusable knowledge will need to be used by *this* project if there are any problems that need to be solved. So, not only is this Causal Mapping methodology reducing the number of problems you need to solve, it is also building reusable knowledge that you can use to more efficiently solve the problems that remain.

The Big Picture: Seeing Your Big Picture

This chapter has focused on the details of Causal Mapping. In understanding those details, we don't want to lose sight of the bigger picture that we are trying to accomplish: enabling your teams to achieve "True North" (to establish that "Success is Assured" prior to making key decisions).

The Causal Map plays a key role in that it connects the detailed issues that must not be forgotten to the big picture that you are trying to accomplish! The big picture is represented by the customer interest decisions marked with red target symbols; those are the key objectives you are trying to satisfy. The stuff in between connecting those customer interests are the critical details that you need to know in order to satisfy those objectives, including the decisions you can directly control, the knowledge you need to know to make those decisions, and the knowledge gaps that you currently have that need to be closed before you can make them (highlighted with question marks).

In other words, the Causal Map serves as *your* big picture of the situation!

In contrast, the first big picture is typically the integrated project plan or schedule; but that rarely makes visible more than just the high-level events and major dependencies – the technical whys and causal dependencies are not exposed – and thus many of the knowledge gaps will not be exposed. The first big picture with any technical depth may not appear until the initial full-scale prototype is designed for simulation or assembled for detailed testing; that is far too late to avoid expensive rework. And the knowledge gaps may not be clearly visible even then, often being covered up by assumptions made in getting to that point design.

So, when a leader asks his or her team for a Causal Map in a K-Brief, it is not a request for more paperwork; it is a request for the big picture to allow the leader to coordinate what needs to be done to make the right decisions. It exposes enough technical detail to identify the required knowledge and uncover the knowledge gaps.

However, we sometimes get pushback that the request still sounds like a request for yet another report, and that can be problematic in organizations committed to reducing the number of reports and other paperwork burdens on their teams. The key to making the distinction clear is in how the leader uses that K-Brief and Causal Map.

If the leader listens to the presented K-Brief and Causal Map as if it was any other slide set presentation, not digging into the details but simply checking off that it was done, then, yes, you are just recasting paperwork in a new form.

If, in contrast, the leader is checking that the big picture objectives match the real objectives of the effort, is challenging the knowledge connected to those big picture details, and is looking for alternative resources and approaches to more efficiently close the identified knowledge gaps, then what is being asked for is not simply a different format paperwork but rather a visual tool for managing the team's efforts.

When the teams see their leaders using the K-Briefs and Causal Maps in that way, they will not only be content to provide it, they will be motivated to provide sufficient detail for the leaders to understand. In other words, the teams will come to realize that they play a critical role in providing the visibility those leaders need to make complex decisions based on the knowledge in the larger organization.

Beyond mere scheduling, Causal Maps and their associated K-Briefs can eliminate much of the conflict caused by different priorities and different expertise. At Teledyne Benthos, for example, engineering and marketing were in a constant state of conflict, marketing was demanding all requirements be met, and engineering was complaining about the infeasibility. When the company leadership demanded that such discussions were to be held around a K-Brief containing the relevant visual models, the heated conflict ended and the teams began to work together productively to close the knowledge gaps and find the feasible parts of the design space. That dramatic change from enemies to efficient collaborators was directly tied to the use of visual models.

You may be surprised at the many dramatic benefits gained from having a big picture to Lean on that focuses the team on the key decisions that need to be made and the knowledge required to make those decisions, and thus the knowledge gaps that need to be closed prior to making those decisions.

Software Tools

Software is not needed for Causal Mapping. Causal Maps should just be how you take notes, how you capture what's being said in meetings on a whiteboard, and so on. However, software tools, if designed to support rapid learning, ideation, and decision-making, can be hugely beneficial. And if designed specifically to leverage Decision Maps, they can be hugely powerful in automating things like Trade-Off Charts and decision optimization. We don't want this book to be about software tools, nor do we want people to delay acting on and benefiting from Causal Maps waiting on software tools, but we also don't want to fail to mention that there are tools out there, if you are interested in such. See www.TargetedConvergence.com for our Success Assured® software tools, specifically designed for Rapid Learning, Causal Mapping, and Trade-Off Analysis based on reusable Decision Maps, Multi-Dimensional Set-Based Decision-Making, Eliminate-the-Weak Optimization, and so on. So, while software tools are not enablers of Causal Maps (just accelerators), they are enablers of some of the more advanced multi-dimensional Set-Based analyses that can drive additional benefits.

Chapter 5

Aligning "True North" across Organizational Boundaries

Imagine an armada of ships organizing to sail to a developing country dealing with a natural disaster. Workers are needed as well as heavy equipment, medical supplies, and doctors. They all must be coordinated to get to the same location at the same time and immediately be productive. The ships are from different countries, leaving from different ports of call, and the types of ships sailing are different. Some can go very fast, others have better navigational equipment, and some are very old and rudimentary. Some ships may be delivering supplies to other ships. All have the medical or support skills to deal with the situation; however, they may each have their own additional conflicting priorities. How do all the ships get to the location quickly and coordinate efforts to resolve the situation? What if a storm comes up? What if conflict occurs? Achieving their goal becomes harder and more complex. How is all this organized and synchronized? They will need more than a compass that simply points north; establishing that "Success is Assured" will involve ongoing coordination of their individual objectives such that they together achieve the larger goal.

Over the years, we have watched situations like this unfold and have seen many roadblocks that have delayed efforts and unfortunately extended the effects of the disasters. Eventually, the situations are resolved, but seldom smoothly, and we all know such poorly coordinated efforts will happen again. Even if the leader of the natural disaster team used all the enablers defined in Chapter 3, keeping all the teams focused on the key objectives would be difficult during the chaos. Rapid learning and decision-making must be fast and effective in each of the sub-organizations but still remain coordinated. The ultimate goal of saving those affected has to be the focus of all the teams in this situation, despite each having their own conflicting priorities.

While certainly not as dramatic or with the same consequences, complex product development faces similar challenges. Knowledge gaps are everywhere. What does the customer really want? What can we do and when? What can our suppliers do and when? How do we get started? How do we ensure success-ful coordination across subsystems? There is an old story about a Toyota Chief Engineer that had just finished examining a brand new model from a competitor. He stated that all the subsystems in the car were world class; but the car itself was not good. His conclusion was that they must have had a lousy Chief Engineer. At Toyota, that is clearly the role of the Chief Engineer; at other companies, it could be the lead systems engineer or the product owner or even the president of the company. In many companies, it is more of a committee approach. Regardless of the form of leadership, it is critically important for any complex process to keep the decision-making at all levels aligned properly to the larger system objectives.

In the prior chapters, we laid out enablers for rapid collaborative learning and effective Set-Based concurrent decision-making. The goal is to find or create knowledge that establishes that "Success is Assured" prior to making critical decisions. That gives the global array of decision-makers a consistent compass and teams that can act accordingly; but there is still need for ongoing focused coordination across organizational boundaries to ensure the optimum solution is delivered to the customer.

In this chapter, we will introduce some simple practices that enable ongoing realignment of the definition of "True North" across the corporate and organizational boundaries. In other words, they will enable the synchronization of decisions and learning priorities throughout the Decision-Focused Learning process until establishing "Success is Assured" prior to the key specification decisions being nailed down.

The Requirements Management Process Must Change

The traditional process is to establish and manage hard requirements and specifications from the start of the project. But that is like giving each ship's captain precise instructions for the entire mission as they launch, ignoring the possibility of storms, ignoring the possibility that they won't each be capable of carrying out those precise instructions, ignoring the possibility that the disaster situation will be different than expected, and so on. In the fuzzy front end of our development processes, when there is still much to learn, drafting hard specifications involves making critical decisions without adequate knowledge and hoping that those decisions will lead to success. In other words, that traditional solution to coordinating multiple organizations violates our "True North" directive.

At Toyota, there was another fundamental difference that Dr. Allen Ward's team observed: unlike the accepted best practices, Toyota did not nail down requirements and specifications early, but rather just set rough targets and allowed the specifications to emerge along with the rest of the learning. They actively worked with their suppliers in the earliest development stages, closing knowledge gaps and using that learning to evolve from rough targets to more precise specifications. In other words, their Set-Based concurrent engineering had been extended even across their customer and supplier boundaries.

Some may push back that doing such is easier for Toyota since they are targeting markets, not specific customers with specific hard requirements. To win the business, companies may need to accept hard requirements and then design around those accordingly. Where that is necessary, that can be done; but if it is not known that the set of hard requirements is truly feasible, closing that knowledge gap should become the highest of priorities. In contrast, more often, that feasibility is taken as a given: most just think, "Of course we can do that, we agreed to do it in the contract."

Requirements Cannot Be Specified Up Front, They Must Be Learned!

Generally, all projects start with some set of customer requirements for the product. And similarly, sub-projects (subsystems) tend to start with a set of specifications partitioned down from those system requirements. Typically, however, those requirements (or specifications) are not feasible together. That is, you may be able to satisfy any one of them by itself, but it is not possible to satisfy them all at once — at least not with today's technology.

Engineers very often complain about their customers (whether end users, downstream corporate customers, or internal product organizations) specifying requirements that are infeasible. Often, project woes are blamed on such impossible requirements. However, such complaints are not fair: what is feasible or not depends on what the supplier is capable of building; they are the experts in that, not the customer. All the customers know is that they'd really like more of this or a better that; wouldn't we all? They don't want to ask for the impossible any more than their suppliers want to be asked for the impossible. However, they also don't want to ask for less than they can get, if more is feasible. So, all they can do is ask for what they'd really want until the supplier informs them of what is feasible.

So, given customers will generally be specifying somewhat more than is feasible, the supplier will have to make trade-offs in their early decision-making. So, to make those trade-off decisions properly, the suppliers will need some guidance from the customer on how to make the right trade-offs.

A common solution is prioritizing the requirements. However, the relative priorities of different requirements depend heavily on where you are in the design space. For example, until you have enough thrust from the jet engine for the aircraft to take off or perform other essential maneuvers, getting more thrust may be the highest priority. But once you can perform all the essential maneuvers, additional thrust may be of only moderate priority, until the thrust begins to exceed other structural limits of the aircraft — at which point, additional thrust may be zero priority. Asking the customer to fully prioritize all their requirements for all different combinations in the design space would be an unreasonable request. Worse, in complex systems, how the different features combine is what enables the additional capabilities the customer is interested in. So the priorities of many requirements will depend on the values that can be achieved for the other requirements.

That is why the visibility of the design space is so important. Engineers need to make visible the trade-off curves that show the levels of the competing requirements, such that the customer can evaluate where on those trade-off curves they would like to be. If the cost-versus-performance curve is very steep, they may prefer a small drop in performance to achieve a big drop in cost; conversely, if the curve is very shallow, they may prefer to spend the small delta cost in exchange for a large performance gain; if the curve goes from shallow to steep, they may want to move up that curve until it becomes too steep. But where the curve becomes too steep may depend on where the other requirements are being set.

Asking the customer to somehow evaluate and communicate such preferences for all possible shapes of curves and levels is ridiculous. Thus, for complex products, our development processes need to admit that the requirements and specifications must be learned as part of the development process; they cannot be demanded as inputs to that process!

However, the converse is also true: you can't ask the supplier engineering organizations to generate all possible trade-off curves — at least, not in a reasonable time frame. You need to give them guidance on what part of the design space you want to be in, so that they can focus on what they can and cannot do in that design space. Some requirements in that space will be easy to achieve; no trade-offs are necessary. Others will be very difficult; but the more flexibility allowed, the easier it will be.

So what should customers be specifying to their suppliers at the start of the process such that it will give them guidance on where the design needs to be, such that they can identify the key trade-offs that need to be made visible to the customer, such that they can then learn where the customer wants to be on that trade-off curve?

Our Approach: Establishing Targets

Our approach is to specify Targets for each customer interest that consist of two values or levels: the *Goal* level and the *Veto* level.

To understand those two Target levels, consider the chart in Figure 5.1, showing how a customer's satisfaction (on the Y-axis) tends to rise as a customer interest decision is changed from worst to best (on the X-axis). So, top speed might be going from a low speed on the left to a high speed on the right, whereas price would be going from high cost on the left to low cost on the right.

Why is this shape of satisfaction curve common? Typically, there is a level below which the product becomes uninteresting — they would never use it — and thus satisfaction falls to zero. When the customer interest decision is improved enough that the product becomes usable, the satisfaction quickly rises.

The vast majority of value curves look something like this:

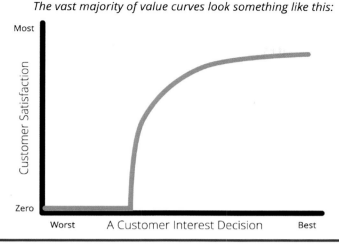

Figure 5.1 Generic customer value curve.

But with yet more capability, satisfaction rises less and less. At some point it may even go flat: further improvement has no real value to the customer.

For example, consider cell phones. Today, most of us would not even consider using the early cell phones that were so large you couldn't even fit them in a briefcase or purse, let alone your pocket. As the phones became smaller and lighter, more people were willing to carry them around in pouches on their belts or in their purses. Once they became small enough to fit in your pocket, most everybody wanted one. But once they were so small and light that you didn't even notice them in your pocket, further reductions in size had little to no value. (In fact, the desire for large screens started pushing phones in the opposite direction; we'll come back to that.)

Of course, not all satisfaction curves will be that shape. And no, we are not going to advise you to ask customers to provide such satisfaction curves for every requirement — or any requirement. Not only is that just as infeasible a request as priority numbers, but worse: by the time we did all the work to collect those curves, they would be out-of-date. As described in the previous paragraph, cell phone customer satisfaction curves have been in constant flux: from phones too large to put on a belt being desirable to phones that need to fit comfortably in a pocket; from flip phones being "in" to flip phones being "old school." Given those curves are constantly moving, they are not worth investing a lot of time in as they will not be reusable.

Instead, we recommend you identify just two key levels of each customer interest (as shown in Figure 5.2) that will largely characterize that satisfaction curve.

1. Where does your interest in this product fall off dramatically? We call this the Veto level (the level at which you would *veto* the product).
2. Where would you be "completely satisfied" such that further improvements are not really too interesting to you? We call this the Goal level (the level at which further improvements are not worth any extra cost).

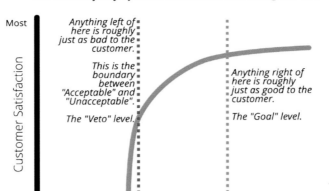

The vast majority of value curves look something like this:

Figure 5.2   Characterizing the value curve with two Target levels: Goal and Veto.

From an engineer's point of view, the design space below Veto can be eliminated as weak right up front. And for the most part, the design space above Goal is uninteresting because it has no additional value but will likely have additional cost. So, if you can get every customer interest to Goal level, then you want to stop optimizing and just ship that product as soon as possible; you have a winner.

More likely, you will be able to get many to Goal level, but some will be competing trade-offs such that you could achieve one or the other but not both. And the best case for the customer might be a little of each. That's where you want to go out and do a little more learning with the customer — just on those critical trade-offs, as shown in Figure 5.3.

In other words, the two simple target levels (Goal and Veto) allow you to quickly eliminate the weak parts of the design space and focus on the regions

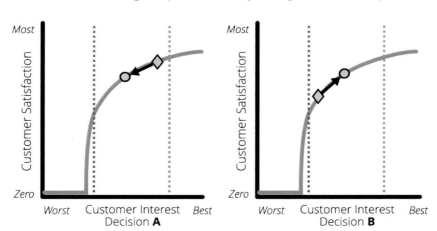

*Customer: "I will give up this much of **A**, to get that much of **B**."*

Figure 5.3   Between Goal and Veto, trade-offs must be made.

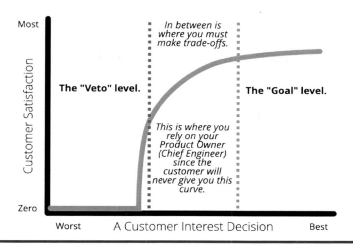

Figure 5.4   Above Goal and below Veto the decisions are easy.

where you need to do a little more learning, because you will never get an accurate satisfaction curve, as summarized in Figure 5.4.

Instead, you need to present the actual trade-offs between the competing customer interests and ask, "How much A will you give up to get how much more B?" or "How much A and C will you give up to get how much more B?" and so on. In other words, they want to see the specific trade-offs between those competing customer interests. For example, Figure 5.5 is the trade-off between noise from a muffler and the Back Pressure it puts on the engine, for a muffler of a specific volume and weight.

As a customer supplying target levels to your suppliers, you are essentially saying, "I am not interested in solutions that don't at least exceed Veto levels of every customer interest; don't bother me with such options. I would prefer you hit all the Goal levels; but if you can't do that, then I'd like to understand what options you can hit between Veto and Goal levels such that I can give you guidance on what would work better for me."

Contrast that with traditional point specifications where, if the supplier cannot hit all the specifications, the customer gives very little guidance on what

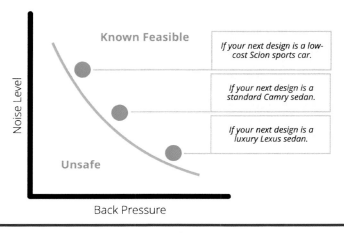

Figure 5.5   When making trade-offs, the specific levels matter.

alternatives they would consider. Adding priorities onto traditional point specifications is not much better because it doesn't bound how low the supplier might go on the lower-priority items.

Further, note that the Veto levels can be treated as feasibility constraints, making them an accelerator. In other words, you can very efficiently use the Veto levels to eliminate the weak parts of the design space. As long as some design space remains, you can then focus on the harder trade-offs. If no design space remains, then you can very quickly get back to the customer and let them know that their Veto levels are altogether infeasible and confirm that those are truly their Veto levels.

We often get pushback on setting these two Target levels (Goal and Veto) when they are interpreted as the more traditional *Ideal* and *Marginally Acceptable* levels. They argue that Marginally Acceptable levels are not acceptable for this project; they want the teams designing for the real goals of the project.

However, our Goal and Veto levels are not just odd renamings of those traditional levels, where Ideal means "the best result a team could hope for" and Marginally Acceptable means "the value that would just barely make the product commercially viable."[1] While identifying the Ideal level can be a very useful technique for idea generation (and one that we promote), that is not what we are trying to accomplish with our Goal Target Level. And we would argue that "barely commercially viable" isn't a particularly interesting Target level and that commercial viability can't really be defined metric by metric. Rather, a customer's willingness to accept a lesser value for one customer interest is highly dependent on the product excelling in other areas. In other words, commercial viability is about the set of trade-offs.

Thus, the intent of our Goal and Veto Levels is around defining the interesting trade-off space in such a way that we can *eliminate the weak* (the uninteresting parts of the trade-off space). The Veto level defines the value of the customer interest where it doesn't matter how good anything else is, the customer would still reject (veto) the product due to that one customer interest being unacceptable. And on the flip side, the Goal level is the point at which the customer interest is good enough that you wouldn't trade off anything else to get any more of that; they attach very little additional value. If you have a design that hits all the Goal levels, then just stop optimizing and ship it as quickly as possible; you have a winner, you're just wasting time and money doing anything more.

So, when faced with such pushback, particularly from the Project Managers or Chief Engineers, we respond: the Veto level is **not** a level that you are saying is "acceptable" up front; that is actually the Goal level. For levels below the Goal levels, you are saying, "You'll need to come show me the trade-offs

---

[1] Karl T. Ulrich and Stephen D. Eppinger, *Product Design and Development*, New York, NY: McGraw-Hill, 2016.

that need to be made, and I'll give you feedback on what trade-offs I consider acceptable." With the Veto levels, you are saying, "Don't even bother me with levels below this; there is no set of trade-offs of the other customer interests that I would ever accept below this level." You are eliminating the weak right up front. So don't be overly aggressive with those Veto levels, because no analysis or learning will be done in that part of the design space; you are cutting that off right up front. Major rework will result if you set those Veto levels too aggressively.

With that explanation, we are then able to get the customers/product managers to set good Goal and Veto levels, where they had historically been very resistant to setting good Ideal and Marginally Acceptable levels.

The Critical Role of Chief Engineer or Product Owner

For complex products, even with Goal and Veto Target levels providing much guidance, there still may be too many trade-off decisions to be learned to resort to LAMDA Discussion with the customer on each. That is where the *customer representative* role of the Toyota-style Chief Engineer or the Agile-style Product Owner becomes critical. They are tasked with understanding the customer well enough that they can make many of these trade-off decisions on behalf of the customer.

To do that well, they must actively learn why the customer is interested in those target levels and any important levels in between. But more importantly, they must learn how the different customer interests will be used together to accomplish whatever it is the customer wants to accomplish.

When trying to develop a more competitive minivan, the Toyota Chief Engineer left Japan and drove minivans more than 53,000 miles all across North America in order to experience his product in its target environment[2] (LAMDA Look). Similarly, the Chief Engineer of the Avalon lived with a family in California for a month.[3] That is the level of understanding we are calling on the Chief Engineers or Product Owners to achieve in order to make those critical trade-off decisions properly; otherwise, you will need to actively involve the customer in making those decisions.

Done right, that customer learning can be fairly reusable if it is continuously improved. It is not knowledge about a specific set of Target levels but rather the *whys* underpinning those Target levels. Capturing such understanding with visual models in a K-Brief can facilitate broader understanding among your team and the desired continuous improvement.

---

[2] Nina Padgett-Russin, Toyota takes to road to improve new Sienna, *Chicago Sun-Times*, February 24, 2003.

[3] Allen C. Ward and Durward K. Sobek II, *Lean Product and Process Development*, Second Edition, Cambridge, MA: Lean Enterprise Institute, 2014, p. 118.

The Critical Role of Set-Based

Point-based product development makes coordination across suppliers and subsystems an intractable coordination problem, as many who live in that world know all too well. The shift to Set-Based development allows the teams to establish "Success is Assured" for sets of possibilities. As those sets narrow, the coordination is simply to make sure the sets on each side of each boundary still overlap.

Similarly, in a world where each team is trying to "pick the best" up front, with "best" defined from their own business priorities, the odds of those independently chosen "bests" aligning is horribly low. If instead the teams are initially focused on eliminating the weak, eliminating the solutions where "Success is Assured" is unlikely from their perspectives, and then communicating what is being eliminated, then those teams can converge to better parts of the design space together.

However, making the shift from pick-the-best hard specifications to eliminating the weak as you learn via rough targets can be extremely difficult without executive action. Hard specifications have become integral to how purchasing organizations manage suppliers, how contracts are negotiated, how government bidding is kept "fair," how regulators judge processes, how organizations are certified, and so on. Those issues will be discussed further in Chapter 6; in this chapter, we just want to focus on how it should work and the resulting benefits.

Once "Success is Assured" has been established, they can then continue that collaborative convergence to find more optimal parts of the remaining design space, moving steadily toward "best" but checking the coordination each step of the way (checking that their sets still overlap with each narrowing decision).

The net effect is a massive decoupling of teams such that they can operate more independently. In many product development organizations, the long cycle times are dominated by "waiting": one team waiting on the results from another team so that they know the "point" the other team is deciding on, such that they can proceed with their own point-based work. Many organizations that adopt Set-Based practices will point to the concurrency and reduction of wait time as the major source of speed-up of their processes.

Scheduling Integrating Events

In Chapter 3, we introduced Integrating Events as a mechanism for integrating the team around the narrowing of a set of decisions, thereby pulling the knowledge necessary to establish that "Success is Assured" for those narrower decisions. Integrating Events become an even more critical enabler as you extend these practices across the customer–supplier organizational boundaries.

We need to discuss how to schedule Integrating Events. For a single team, there is a fair bit of leeway. Across teams, a good sequence for making decisions and appropriate timing can be learned over time. But when first trying to coordinate learning and decision-making across customer–supplier boundaries, a more structured approach is necessary.

Given that competitive products are continuously improving, the design of a product is highly dependent on its target release date. We recommend that a program adopting "Success is Assured" decision-making practices start by backward scheduling the back end of their product development process from that target release date largely as they normally would (Figure 5.6). Identify the long lead time items to determine when specifications need to be nailed down. Identify key dependencies to drive when other key decisions need to be made. In engineer-to-order contracts there may be key gates whose dates have been (or are being) negotiated. Start the process by identifying all such key decisions dates, assuming the back-end process will proceed with the normal optimism.

Then, for each identified decision date, ask what needs to be known to establish that "Success is Assured" prior to making that set of decisions. That analysis should be relying heavily on the separately developed Causal Map of the key customer interests, the dependencies and limits, and the related knowledge gaps.

At that point, there may be nearer-term decision points that appear particularly challenging to close all the knowledge gaps required to establish that "Success is Assured." For those, start to challenge to what degree those decisions need to be narrowed at that point. Leverage Set-Based practices to try to relax the decisions enough that some of the knowledge gaps can be closed later, or perhaps just closed for certain levels of "worst case."

In any case, do not try to continue the backward scheduling into the learning tasks. By their nature, learning and innovation cannot be scheduled accurately. Further, to maximize efficiency, teams should seek to leverage the learning each step of the way to reduce the amount of additional learning that remains to be

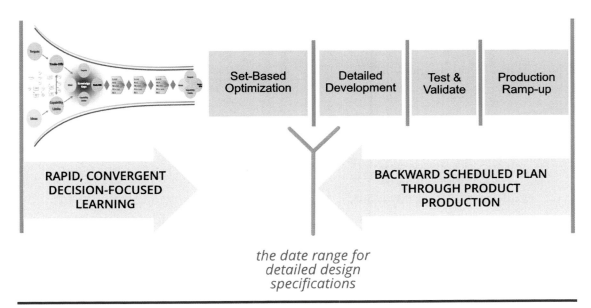

the date range for
detailed design
specifications

Figure 5.6   You can't backward schedule the learning, so move the learning up front.

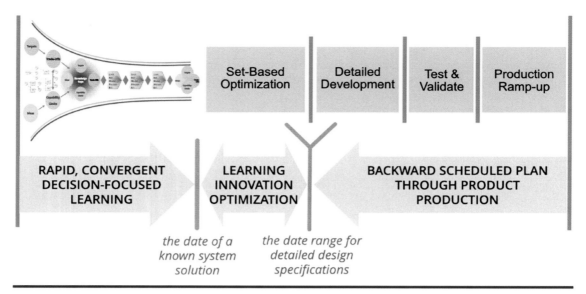

Figure 5.7 The optimization period can be a pad if the learning takes too long.

done. Premature scheduling can and almost invariably will get in the way of such efficient convergent Decision-Focused Learning.

Instead, embrace a front-end process made up of focused learning *sprints*. At the start of each sprint, identify what learning will potentially eliminate the most design space as weak, and thereby eliminate the need to close as many of the other knowledge gaps as possible. At the end of each learning sprint, ask the impacts of what was learned. Where should be the focus of the next learning sprint?

For each scheduled Integrating Event, "Success is Assured" is ideally established well enough in advance that additional learning can be done to further optimize — to further narrow the design space to the more desirable parts of that design space. But that must be done in the context of the other decisions that remain ranges of possible values — that have not been fully narrowed.

That potential optimization time (as shown in Figure 5.7) serves as a time buffer: if the learning to establish that "Success is Assured" for that set of decisions ends up taking longer, then that effort can continue eating into the optimization time.

What if the Learning Takes Too Long?

If we eat through all of the optimization time, such that we reach the decision point but have not yet established that "Success is Assured" for that set of decisions, then the team has an additional choice to make.

- Delay the decision point by some time period to give more time to close the knowledge gaps but thereby delaying the whole program by that same amount.

■ Fall back to traditional decision-making practices: guess the right answer and manage the risk that you will have additional rework later.

Even if the latter is chosen, all is not lost; that is not at all a fall back to status quo. Let us explain...

First, given experienced engineers, our engineering intuition tends to serve us well when issues are isolated. It is the unexpected interaction of numerous knowledge gaps in complex product situations that tend to burn us. Thus, as we close knowledge gaps, we tend to reduce the likelihood of rework exponentially. So, if that up-front effort closes 50% of the knowledge gaps, it may reduce the rework 80%–90%. If that up-front effort closes 70% of the knowledge gaps, it may reduce the rework 90%–95%.

Second, as we close those knowledge gaps in a Set-Based way, we are building knowledge of the larger design space, such that when a surprise occurs and a decision must be changed, we know most of the impacts of such a change. That often allows us to deal with the surprises much more quickly and with far less risk of negative impacts on the rest of the design (and follow-on surprises and rework).

Finally, there is the benefit of constructing reusable knowledge for the next project — though not all programs consider that much of a benefit.

Mentoring and Cadence for the Big Picture

In Chapter 3, "The Enablers," after introducing Decision-Based Scheduling and Integrating Events, we discussed the importance of Mentoring and Cadence, which are really two separate enablers that are each of great importance but which operate hand-in-hand. Those same two enablers become even more important as we extend Decision-Based Scheduling and Integrating Events across organizational boundaries.

Within a single organization, with the manager or technical lead in the role of mentor, that mentor can generally ensure that all new learning and all convergence decisions are properly communicated. Alternatively, a regular cadence (e.g., once a week, three times a week, daily) of quick stand-up meetings where new learning and convergence decisions are reported to all can efficiently ensure that everybody is kept equally informed.

Across multiple organizations, those simple mechanisms quickly become infeasible. So, how do we ensure that everybody is kept equally informed in the face of ongoing learning driving rapid decision convergence?

One alternative is a regular cadence of meetings bringing together the managers or technical leads of the various sub-organizations where they report the learning in their sub-organizations that may impact others and where they report any convergence of key decisions. For moderate-sized organizations, these can be frequent enough and comprehensive enough that these will work well. They are a natural extension of the daily stand-ups and thus are a natural fit.

As organizations and supplier networks grow larger or more complex, however, such meetings may become too big or too infrequent to be adequate. As learning occurs or decisions are converged, that needs to be immediately communicated to any whose work may be impacted so that they can verify that feasible design space remains and refocus their learning efforts on that remaining design space. Any delays in that communication can result in inefficiency and rework (not to mention frustration and conflict). If adequately frequent meetings are not feasible, then an active mechanism will need to be put in place.

For moderate-complexity products, that active mechanism can be the Entrepreneurial System Designer role (aka, the Toyota-style Chief Engineer or the Agile-style Product Owner). They can play the role of Mentor on all interface decisions (decisions that cross organizations). In that Mentoring role, they see all the learning and all the convergence of those decisions and can actively communicate that to the other organizations and verify that feasible design space remains. They can use their engineering judgment regarding how urgently different learning or convergence needs to be communicated, and thus they can actively manage the cadence as appropriate.

As product complexity increases, a single person will not be able to mentor all of that. Toyota's solution is to place Junior Chief Engineers under the Chief Engineer, each with their own focus, mentoring some subset of the interface decisions, communicating some subset of the learning and convergence. The Chief Engineer then mentors the Junior Chief Engineers to ensure all the necessary collaboration is happening.

In scaling Agile methods, *scrums of scrums* and *agile release trains* have been proposed. In project management, numerous scaling mechanisms have been used. Many other variations are possible, including combinations of the above. The right solution will depend heavily on the nature of the organizations themselves, so we will not be prescribing a particular solution, but we will suggest some criteria for evaluating potential solutions.

First and foremost, you need to make visible the key decisions, which of those are customer interests, and which of those are interface decisions (decisions shared across organizations). You then need to make visible your knowledge gaps regarding those decisions and how they are related (how they impact each other). That is the "big picture" that every sub-organization needs to remain aligned with.

With the visibility of that big picture and the timing information established by scheduling the Integrating Events, as discussed in the previous section, you now can judge the necessary cadence for communication across organizations regarding that learning and convergence of decisions. For example, if the next sequence of Integrating Events is four weeks away, you may need a cadence of a few days to a week; whereas, for Integrating Events that are four months away, a cadence of a few weeks may do just fine. Any mechanism that can deliver that cadence, whether regularly scheduled stand-up meetings or active management by someone or otherwise, may work just fine. But if that mechanism cannot support the required cadence, you should look for alternatives.

That big picture, the sequence of Integrating Events, and the mechanisms to support the Mentoring and Cadence will together form your system-level decision-making best practices. Whether you manage them with Project K-Briefs, Visual Management boards, *obeya* rooms, and so on, you should be able to continuously improve them such that they become reusable knowledge in the form of tools you can apply to future similar efforts.

For This Program, We Do Not Care about Knowledge Reuse!

We often hear: "This program is a bit of a one-off"; "We don't expect to be doing another product like this for a decade, if ever"; "The time frame for potential knowledge reuse is not of interest to our management; they will tell us not to compromise the program timing at all for potential future knowledge reuse benefits."

Given that most Knowledge Management initiatives have resulted in systems where the knowledge goes in but very rarely comes out, getting such feedback from programs is not at all surprising — even in companies with corporate initiatives to improve their Knowledge Management.

That makes this a good time to emphasize a key point regarding knowledge reuse:

> **Knowledge reuse is *not* just for future programs, it is for *this* program!**

Consider the previous few sections: in the case there is a surprise and a decision needs to be remade, the amount of resulting rework may be dramatically reduced by the fact that you have built reusable knowledge regarding the surrounding design space.

Consider that as we go through the sequence of Integrating Events, we are continuing to narrow the decisions that were narrowed in prior Integrating Events and related decisions. In other words, even without surprises, we will tend to reuse the growing body of knowledge as we continue to refine the decisions we are converging.

Consider that as we do the learning regarding complex products, we need to pull in expertise from different areas. Each time we do, we need to be able to present the existing body of knowledge to get their feedback on what is missing.

Consider that as we continue to learn the complex multi-dimensional design space, engineers may come up with different innovations that may result in new alternatives to be evaluated. Those evaluations should reuse that growing body of knowledge about the best way to make those interrelated decisions.

Don't consider investment in reusable knowledge just an investment in future programs; it is an investment in the future of *this* program! Consider it a form of Risk Management. The ideal Risk Management will close the knowledge gap and

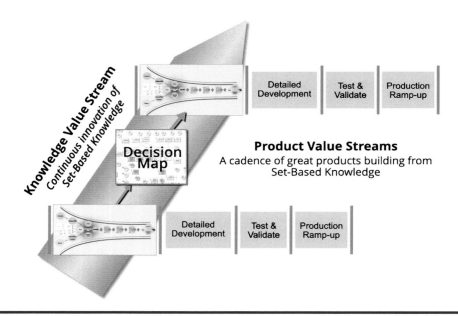

Figure 5.8   Learning-First Product Development process.

eliminate the risk. The next-best Risk Management will build up knowledge of the design space around the chosen point so as to minimize the impact of the risk that the decision is wrong. The worst form of Risk Management will simply add time padding for the inevitable rework should that risk occur.

Even if your management does not adopt the Toyota-like attitude that "Product Development is not the process of designing cars; it is the process of developing knowledge about cars. Car designs will emerge as a by-product", as depicted by the *Knowledge Value Stream* graphic in Figure 5.8, we would still encourage them to embrace the development of reusable Set-Based knowledge of their larger design space for the huge benefits it will have on the current program, both in improved decision-making and in quicker and more effective reaction to surprises. And in the context of working with your customers and suppliers as part of a larger supply chain, it will help you better influence the decisions to be made and better manage surprises from decisions that have been made beyond your corporate boundaries.

A Rapid Cadence of Successful Products

A rapid cadence of consistently successful products can become the norm for organizations willing to invest in establishing a Knowledge Value Stream of reusable Set-Based knowledge. With that in hand, the decision-makers in the front end of each *Product Value Stream* can demand that they have the knowledge to establish that "Success is Assured" prior to making key decisions.

We believe you now have all the ingredients necessary to establish such a Knowledge Value Stream and feed the early decisions of your Product Value Stream. Chapter 2 gave your decision-makers a "True North" to guide their decisions based on knowledge. Chapters 3 and 4 gave your engineers (or more generally, your knowledge workers) the enablers to deliver the knowledge required by your decision-makers. This chapter has given simple techniques to coordinate those efforts, even across organizational boundaries, with the Causal Map serving as the big picture guiding those efforts.

However, when crossing organizational boundaries, there may be additional structural hurdles that must be overcome: contracts, regulations, communication structures, information technology incompatibilities, program management structures, and rewards and incentive systems, among others. Those can be addressed following the pattern laid out in Chapter 2: by consistently applying "True North" to each. But that must be done proactively, and it will take strong focused leadership to make that shift possible.

Chapter 6 will outline some strategies for making the shift from your existing practices to these new practices. And Part II of this book (Chapters 7 through 16) will illustrate those key ingredients on a fictional (but realistically complex) project such that you can experience putting some of that together, prior to adopting the practices in your own organization.

Chapter 6

Making the Transition

The world of product development is a tough place to work.

- Problems on prior projects consume the limited resources, making it difficult to get work done on new projects.
- New projects begin with a seemingly infeasible and inflexible set of requirements, along with a fixed schedule and budget.
- Internal processes, reviews, and expectations often seem to present more barriers than they provide help.
- Suppliers with unreasonably long lead times force premature specifications.
- There never seems to be enough time and resources to do what you know needs to be done the right way.

With all these challenges, just the idea of making some significant changes to the front end of the product development process can feel overwhelming — even risky. There is a strong tendency to stick with what you know, hoping that a few minor tweaks and a little more diligence will produce much better results.

However, if you have read this far in this book, then you have come to believe, perhaps uncomfortably so, that there are some fundamental problems early in your product development process — problems that need to be addressed at the root cause.

For organizations that are really struggling in product development, initiating a change to "True North" may be a no-brainer. After all, how much worse could it (or anything) be than what they are doing now? At the other end of the spectrum are the innovators and industry leaders. The logic of establishing that "Success is Assured" early appeals to these organizations, and they usually have the resources and leadership to initiate a systematic evaluation and implementation of the Enablers.

Most organizations fall in between. They have recognized problems in product development, but executing the current project seems more urgent than fundamentally improving the development process. Their business model seems to be working in that they remain competitive in their market. It may be good enough until their competitors get their product development act together.

And there's the rub. An organization that is not only able to speed their time to market but effectively reuse what they learned on the next project can quickly attain a dominant competitive position. It is like a runner at a track meet that is allowed to start each new race with a lead equal to the length by which they won the last race. Within just a few races, it might become impossible to catch that competitor.

Whichever of these situations your organization is in, we have seen three general strategies for implementing the "True North" principles and practices succeed.

1. Establish "Success is Assured" when Problem-Solving.
2. Identify and Focus on Knowledge Gaps Early on in Targeted Programs.
3. Establish "Success is Assured" in appropriate Pilot Projects.

In this chapter, we will offer you some practical advice on implementing each of these strategies in your organization, including how to address the obstacles that are commonly encountered.

Strategy 1: Establish "Success is Assured" when Problem-Solving

A survey by the Automotive Industry Action Group (AIAG)[1] revealed that the number-one problem facing both AIAG OEMs and their suppliers is poor problem-solving — even though most of those companies had defined Problem-Solving Processes. Further, they identified four key reasons for that poor problem-solving as:

1. Poor root cause analysis
2. Management/organizational culture issues
3. Feeling rushed
4. Jumping to solutions

Traditional Problem-Solving Processes work well on simple, linear problems, such as

- Why is this machine not meeting its tolerance requirements?
- What is causing repeated assembly issues between two parts in this component?
- Why is it taking so long to release drawings?

---

[1] AIAG in collaboration with Deloitte Consulting, Automotive industry's view on the current state of quality and a strategic path forward, based on Deloitte/AIAG Quality 2020 Survey, p. 5.

Unfortunately, most engineering problems aren't linear, they are complex learning processes. In Chapter 3, we introduced a learning process based on the LAMDA and K-Brief Enablers. In Chapter 4, we explained that root cause analysis can't be simply a 5-Why-type analysis but must include three parts:

1. Identifying the root causes (decisions that you control)
2. Understanding what other customer interests could be impacted by changes to those root causes
3. Making visible the trade-offs you will be making when deciding among the alternative solutions

For companies who spend 65%–75% of their engineering capacity on problem-solving or companies having trouble starting new projects because their people are all still fighting fires on the old projects, focusing the transition on first improving Problem-Solving can be very effective and foundational for establishing a "Success is Assured" engineering environment.

A common approach is to introduce the Problem K-Brief (see Appendix VII) and its Causal Map for Root Cause Analysis and mandate that such be used for all future problem-solving efforts. The common failure mode, however, is that people use the K-Brief as a final report format. It is critical to teach the Problem K-Brief not as a form to be filled out but rather as a vehicle for ongoing mentoring between the problem-solver and his mentor (typically his manager). It should be treated as a tool for organizing the visual models needed to facilitate all discussions regarding the problem. And those should be LAMDA Discussions; LAMDA should be integral to the process.

The approval process for each Problem K-Brief should involve showing that the knowledge establishes that "Success is Assured" for the chosen remedy or remedies. For more complex problems, mentors can encourage an *eliminate the weak* process of narrowing the design space. They can also encourage that the learning activities capture reusable Set-Based knowledge rather than simply point-based tests, wherever timing is not impacted or the time can be afforded. That addition can establish some really good habits.

By establishing that "Success is Assured" and thereby preventing the solution to one problem causing other problems, you can also see your Problem Backlog rapidly shrink. Clients we have worked with have had overwhelming problem backlogs that have been completely emptied a year later. That frees up capacity to apply those problem-solving skills to the front ends of the projects.

However, don't forget that there are additional skills to be learned when switching from problem-solving to projects: new types of K-Briefs, new visual models, and new knowledge gap and trade-off identification practices. Treating a project as a product design "problem" won't yield nearly the same benefits.

Strategy 2: Identify and Focus on Knowledge Gaps Early in Targeted Projects

Based on our experience, we suspect that many of your current projects have already made key decisions and are marching (probably racing) toward detailed design with both known and unknown knowledge gaps, or they are already well into detailed design with those knowledge gaps. The result will be design loop-backs and unplanned rework, often very late in the project.

What if you intervened to proactively expose the knowledge gaps and resolve them before committing to (further) detailed design? A combination of LAMDA and Causal Mapping will surface most of the key trade-offs and concerns such that they can be addressed immediately. However, to have a significant effect on the program, it is key that management continues to ask the status of those knowledge gaps, otherwise they are "out of sight, out of mind," and the gaps are soon forgotten.

For companies embarking on establishing that "Success is Assured," our experience has shown it wise to initially develop a team of "Success is Assured" Mentors (SAMs) who can facilitate LAMDA, K-Briefs, and Causal Mapping on the focused projects. Those SAMs can greatly reduce the time to positive impact on those projects, avoiding the delays that would be caused by the teams going through the training and trying to ramp up on their own. Building momentum is key for projects under the typical time pressures.

The next step up beyond just LAMDA and Causal Mapping is to develop a Project K-Brief that lays out

1. The key Project objectives and the key Business Interests
2. The Customer Opportunities being pursued and key Customer Interests
3. The Competitive products and where they stand on key Customer Interests
4. Causal Map(s) for satisfying those key Customer interests
5. A "When will Who do What?" table showing the learning tasks identified to close the Knowledge Gaps identified in any of these

The first three tend to identify the key trade-offs that need to be mapped out in the Causal Map, and the five together surface and close the knowledge gaps more effectively than the Causal Map alone.

Without the pull of trying to establish "Success is Assured" prior to decision-making, the decisions tend to be made per the original schedule (schedule-based decision-making), and the knowledge gaps that are more difficult to close tend to just be accepted as unknowns to be resolved through Risk Management. However, having a more complete list of knowledge gaps plus visibility to their potential impacts and interactions tends to result in more robust Risk Management as the risks are far better understood.

The bigger impacts depend heavily on the degree to which the project management tends to push for those knowledge gaps to be closed, and by that we mean the degree to which they make time for and allocate resources to the closing of the identified knowledge gaps.

Note that this strategy can be employed even on projects or programs that have already moved into detailed design, having already made the decisions without establishing that "Success is Assured." Even then, the sooner the knowledge gaps are identified, the sooner the learning can be accelerated, the less rework that may be caused by that learning, and the more flexibility you may have in working around the learning. Remember P&W's use of causal mapping to eliminate a weight issue at the end of their preliminary design phase?

Strategy 3: Establish "Success is Assured" in Appropriate Pilot Projects

Given you have one or more projects or programs in the fuzzy front end or soon will have, there will certainly be a great desire to avoid the typical levels of engineering rework. This should justify investing some time up front to learning and deploying a new process as long as you feel it will be effective, and nothing will give a clearer answer for how big the benefits might be on such projects.

However, for it to be a true pilot of the methodologies, it is critical that you get the full project management on board, so training needs to start with them. They need to learn what they will be asking their engineers to do, because only then will they be able to take on the role of mentor, and only then will they know the right things to be asking for when mentoring.

Similarly, it is important to get all the key decision-makers on board, even if they are not part of the project management: anybody who will be deciding priorities of what to work on or making design decisions on the product. It is important that all such decision-makers are demanding "True North" in their decision-making.

As part of that training, they can begin working on the development of a Project K-Brief to pull the knowledge identification, as described in Strategy 2. But then, for each of those identified knowledge gaps, they have a problem to solve (how to close that knowledge gap). Those knowledge gaps can then be used to train the teams on establishing "Success is Assured" when problem-solving, as described in Strategy 1.

Beyond deploying both Strategies 1 and 2 on the front end of the project or program, piloting the methodologies will quickly get to points in the process where key decisions would traditionally be made with incomplete knowledge. At that point, they should start employing Set-Based practices more broadly than either of the first two Strategies would do. Training in Set-Based practices can be incorporated into that training, and it is important that it is done as early as possible as it is the key enabler in the fuzzy front end. However, as the need for converging decisions approaches, the development of the required Trade-Off Charts will typically involve some new skill sets in dealing with complex, multi-dimensional problems.

One common failure mode in such pilot project efforts is short-changing the up-front investment in training the managers, the decision-makers, and the engineers. That investment is quite small compared with the overall costs of executing the project. Preventing even a single loopback and the associated rework will typically more than pay for that training time and cost. The example in Part II of this book will illustrate how quickly a team can ramp up. And yet, still the most common failure mode is to shortcut that training at every turn. It is far too easy for people uncomfortable with the new practices to fall back into the old, either consciously or unconsciously. Give your teams the training they need.

Another common failure mode in such pilot project efforts is not managing the naysayers. In a large project, some of the managers and decision-makers will not buy into the new practices and will resist. If they do not provide the required mentoring to the people in their teams, then those teams can end up undermining the transition. Or even more subtly, if they simply keep insisting on the old deliverables and practices such that there is no time for the new, they can undermine the transition. So, it is critical that there is leadership higher in the management hierarchy that is both adequately supportive of the transition and adequately trained and engaged such that they can keep the naysayers in line or be prepared to move them out of the pilot project and into a project still operating with the traditional practices they prefer.

Common Obstacles to the Transition

Regardless of what strategy an organization chooses for its transition, there are a few obstacles that typically stand in the way that all need to actively identify and remediate but may require involvement from higher leadership to be adequately resolved.

Artificial Schedule Pressures

Often, these are motivated by improvement initiatives or ongoing experience with late-process learning. The thinking is that leadership needs to push the organization into the later phases to get to the firefighting (learning) as soon as possible; but in fact, when complexity is high, they are creating very long, slow learning cycles. That pressure needs to be redirected to driving knowledge gap identification and high pressure to find innovative ways to close those knowledge gaps early.

Negotiated Schedule and Budget Pressures

Although the ideal may be open discussions with customers on the knowledge gaps and the closing of those gaps being designed into the agreements, the reality is that organizations still have to compete in a world where your competition and your customers may be willing to bet on guesses; in such a world, your organization may be forced to do the same to win the business. Once done, those guesses become critical objectives, but you must resist them becoming accepted as immutable facts; rather, they should be identified as the most critical of all knowledge gaps. If they need to be changed, the sooner you know that, and how much they need to change, the better you can manage your way to that change.

Negotiated Micro-Management

Another flavor of negotiated obstacles is common in cost-plus types of contracts, where the customer wants a level of control over what they are spending. Such ongoing cost management mechanisms often dictate traditional point-based development practices and can get in the way of early knowledge gap identification and closing that may rightly involve Set-Based analysis. Set-Based analysis can seem to traditional thinking as investment in broader understanding than needed for the current project. Trying to explain that the development of such reusable knowledge is for *this* project, not just future projects, can be a challenge if the customer has not adopted these practices. Classifying front-end learning as overhead can quickly exceed allowed overhead limits. In contrast, classifying late process firefighting as project-specific is no problem, even if that costs 100 times what the up-front Set-Based analysis would have cost. Visual knowledge making the knowledge gaps and the likely cost of rework clear can often overcome this obstacle, but the supplier organization has to be willing to expose those knowledge gaps to their customers. If your organization has this issue, then it almost surely has the next one.

Lack of Charge Numbers for Learning and Set-Based Practices

Whether operating cost-plus or not, a common mechanism to control expenses in the bulk of product development is establishing charge numbers, such that people won't do anything unless they know what number they can charge that work to. For government contractors, there are often legal implications that come along with the use of these. It is very difficult to conduct the necessary Causal Mapping or LAMDA discussions if you cannot get the experts in the room because you don't have a charge number. As noted in the prior paragraph, doing such learning as part of problem-solving at the ends of projects is not a problem: that is classified as problem-solving and can be charged to whatever it is you are fixing. But even though doing that learning Set-Based in the fuzzy front end may be 100 times or more cheaper than doing that learning late during problem-solving (due to the associated rework), such up-front learning and Set-Based activity is often thought of as not project-specific and thus not allowed to use the project charge numbers. That alone can kill the transition to these practices. If you want to do the learning required to establish "Success is Assured" in the front end, then you must clearly communicate your organization's approach to charge numbers for that work. Either make it clear that building Set-Based knowledge of the larger design space in which this program is expected to land can use the program charge numbers or provide charge numbers specific to such early learning. If that approach is not clear to everyone in your organization, then you can safely assume your organization will be spending very little time on establishing "Success is Assured." And do not fall into the trap of "We'll deal with that when we run into it on a case-by-case basis," as that will slow down the front-end learning tremendously (every time you want to do another LAMDA step, you will have to work through program management to figure out what charge number can be used). You must have a clear approach to such charging laid out up front.

Wishful Thinking

An obstacle to getting learning charge numbers set up can also be a more general obstacle to transition to these practices. Too often, our processes are built on the assumption that we start projects knowing everything we need to know — that we just need to execute. Even when we admit there will be some learning, we hope it will be small and will be handled underneath the major tasks. But leaving such learning for that later timing results in the tremendous inefficiency of the rework. So, even though the cost of learning could have been kept small enough by moving it up front, our back end-oriented policies and procedures prevent us from doing the necessary learning in the front end. If the knowledge is not visible, assume it does not exist. If the knowledge that is visible does not clearly establish "Success is Assured," assume that problem-solving and rework is assured.

Specification-Based Practices

As mentioned in Chapter 5, many organizations' current practices are actually built on hard point-based specifications. Such practices can become tremendous inertia resisting change to the use of rough targets in the fuzzy front end such that the customer and supplier can learn the right specifications together. The best approach may be to completely rethink the purpose and nature of those practices by embracing more cooperative and trusting relationships across those boundaries. But that may be too much to ask initially. A much easier approach is to just allow the required hard specifications to emerge *later* in the development cycle. For example, you may set up the contract with rough targets along with dates when the hard specifications will be nailed down. Certain assurances could be tied to the Veto levels. Beyond that, the contract would need to give an out if the learning does not result in identifying design space where "Success is Assured." That alone may not be enough; for example, in government contracting, very often the bidding process is highly regulated and very much cost-driven. Even outside of government regulation, some industries have very cost-driven bidding such that traditional practices would want the pricing nailed down prior to the learning. Given all the different organizations, all the different process definitions, all the different incentive systems, and so on that may be constructed around the current practices, it may take concerted action by an executive leader to bridge all the obstacles, allowing the required learning to happen prior to nailing down the specifications.

Pressure to Commit Early

This comes from a number of different sources. Some will assert that the root cause of poorly performing projects is management not willing to make a decision or failing to commit resources to the decisions it makes. Some improvement initiatives even promote such early decisiveness from leadership to increase early efficiency. Decisiveness has become a desirable trait in leaders, even if that means sticking with a guess that can be quickly proven to be a bad one. We want to redirect that decisiveness to include deciding that resources need to be committed up front to closing critical knowledge gaps and to keeping sets of possibilities open until those gaps can be closed. The commitment we want to encourage is to finding the shortest path to establishing that "Success is Assured," not to decisions based on intuition (guesses).

Successful Gambler Syndrome

A well-known path to financial ruin is a successful day or two of gambling at the racetrack or at blackjack or poker, leading to growing confidence in the gambler's decision-making abilities (when, in fact, he or she just got lucky). Similar happens to many who place their bets in the stock markets, which is an even more subtle trap, as it is considered to be "investing" rather than "gambling." The same happens within our business organizations: leaders are promoted due to their successful decisions, giving them the confidence to continue making decisions that same way. Many of the leaders that have asked for our help have done so after a project failure, when they realized they had done nothing different in this failure compared with their prior successes and concluded that they in fact had just gotten lucky in their decision-making. That realization can be terrifying — and a fantastic motivator for a better way of decision-making. But given many other leaders may not have yet had that experience, working around overconfident decision-makers in very senior positions can be one of the toughest obstacles to the transition. The clarity of visual knowledge can be a solution, when it is available. But making time to create such knowledge may require intervention by other senior leaders. (One side-benefit of using Integrating Events to create true ownership of a set of decisions across a larger set of decision-makers is that when it goes wrong due to relying on a guess rather than knowledge, a whole group of senior leadership can have that experience together, as they all feel they made those failed decisions.)

Not-Invented-Here (NIH) Syndrome

Many will point to this as a key driver of the lack of knowledge reuse: that engineers would rather invent a new solution than reuse somebody else's solution. In our experience, this is not actually true. Sure, most engineers would like to be the inventor of a great new idea; but even more often, engineers prefer to avoid risk. Fear of being blamed for a disaster tends to be a stronger driver of their decision-making. We would argue that so-called NIH behaviors have more to do with a healthy skepticism for knowledge that was developed for a different purpose, with a different set of assumptions, and with unknown foundations. Often, past successful decisions were made based on guesses, not knowledge; so, the fact that they were successful is of little comfort, since they may have just gotten lucky. The lack of clear visual knowledge that drove the past decisions' success makes that past "knowledge" not particularly reusable to risk-averse engineers (rather than any desire to invent something themselves). The preference for their own past inventions is that they have deeper understanding of the assumptions that went into their own inventions. Visual knowledge that makes clear how the different targets would be impacted by the different decisions and assumptions can largely alleviate such fears.

Lack of Organizational Focus

Many organizations, particularly larger ones, have started initiatives over the years (including those listed in Chapter 2, among others). And most of those initiatives can have significant benefits and can be aligned with "True North," so they should not be a problem, right? There is a limit to how many different improvements an individual can focus on and succeed at. And that limit can be even lower for an organization because of the conflicts created when the individuals are pulling in different directions, all blessed as approved initiatives. It is critical that senior leadership establishes clear priorities for the initiative that can have the biggest impact (e.g., fourfold productivity improvement or twofold reduction in product development times, rather than 20% cost savings). The other initiatives can remain blessed, to be worked on at any opportunity, but always in support of the primary initiative. Senior leadership must provide the organizational focus; if you are not senior leadership, then you should seek out the senior leadership you will need to provide that focus.

Lack of Trusted Metrics and Cost Avoidance Not Being Considered Cost Savings

It is easy to measure the improvement of a process that does the same thing in less time or consuming fewer resources; it is easy to determine the time and/ or cost savings. But how do you measure a process that gives you a better result and avoids future rework? Many organizations will flatly state that you can't treat cost avoidance as cost savings, even though 75% of their costs might be things that could be avoided. How do you put a monetary value on *first-time quality?* In manufacturing, these improvements can be measured more easily because the process is producing the same product over and over; you can make an apples-to-apples comparison. But in product development, we are rarely asked to design the same thing again from the same starting point such that we have a fair comparison. Rather, we are always designing something new, with different challenges and different starting points. As a result, improvement initiatives in product development are often woefully underfunded because they cannot present metrics that top executives consider adequately concrete. If those organizations had 10% higher development costs, that would be pretty wasteful but perhaps justifiable given other similarly sized but measurable improvement opportunities elsewhere. But if it is in fact resulting in 400% higher costs and 200% longer development times, then senior leadership needs to step in and bless alternative metrics, or simply fund the initiatives based on the potential improvements.

Putting the Enablers in place and then keeping a focus on "True North" will consistently drive significant improvements. But to fully benefit, to fully make the transition, senior leadership will likely need to help overcome at least some of the preceding obstacles. So, we hope this section has put everybody on alert to these so that they can be identified and dealt with proactively.

Beyond the Transition

Years ago at a Lean conference in Cincinnati, the president of Toyota North America made this statement about the lessons learned from the Toyota Product Development System:

> *Keep it simple, make it visual, and trust your people to do the right thing.*

Product development today is anything but "simple." Over the years, layers of processes have been developed and redeveloped in improvement attempts; yet despite that, making decisions that don't require rework has remained elusive. Many of the natural responses to that rework have involved trying to nail things down earlier in the front end, but this may have actually made things worse: more pressure to make decisions early and less time to learn what is needed to make those decisions right.

Most product development processes today rival the products they are developing in complexity; in some cases, the product's complexity is no match for the product development process's complexity. And most of that complexity is in fact unnecessary and should be eliminated to maximize the benefits and minimize the waste. However, up to this point, this book has not been calling for that elimination. Why?

Until your people have learned the new simpler practices such that they have proven to themselves and their management that those simpler practices are sufficient, there is far too much pressure to rely on the practices that they are comfortable with. And thus, while building shared experiences with the new practices, it is necessary that they co-exist with the old practices.

However, even during that co-existence, some of the old practices will need to be challenged if they cause violation of "True North." Beyond that, some of the old practices may be wasteful enough that they also should be challenged, particularly if it frees up capacity to perform the new practices.

Management should be careful not to add new deliverables to the burden on their people without actively looking for deliverables they can remove. In particular, for any old deliverables that are largely covered by the content in the new deliverables, management should declare that the new deliverables suffice for the old. Beyond that, if doing both the old and the new is creating complexity or confusion, then management should actively reconcile that and remove complexity one way or another.

Given the level of excess complexity baked into existing processes, management should be intolerant of further complexity and open to options that reduce complexity; they should know that is the direction in which they need to be moving in the future.

Final Thoughts

In this book, we have challenged the underlying assumption that insufficient knowledge exists in the conceptual front end of product development (the fuzzy front end) to make quality decisions and that only after detailed design and testing will the knowledge be sufficient. Chapters 3 and 4 have taught tools and methodologies that can enable teams to do that. And Chapter 5 has shown how to apply those even across complex organizations developing complex products.

Significant benefits are seen in most organizations where the leadership has bought into these tools and methodologies, has trained all their people in them, and is asking for visual knowledge in the form of K-Briefs and Causal Maps. However, they often do not see the near-elimination of rework and the associated fourfold gains in productivity or the cutting of development times in half. The typical cause is that it is not good enough for the leadership to ask their organization to change; the leadership has to change how they make decisions as well. The leadership from top to bottom must adopt "True North."

> *We will adjust our processes, our behaviors, and our expectations to insist that we know the impacts of the decisions we are making with enough clarity that we can make desirable trade-offs rather than committing our limited resources to decisions that will likely change.*

Those are the essential ingredients. Everything else taught in this book is optional, required only as needed for your organization to achieve it. But this "True North" statement cannot just be a war cry — a quotation posted on the conference room walls, coffee mugs, and badge protectors. It must change how the leaders make decisions! If the leaders do not have confidence that the knowledge is sufficient to make quality early decisions, then they will rightly fall back on making rapid decisions to get into detailed design in order to do that learning sooner. But the more common failure is that the engineers are happy to use the new methods to develop the required knowledge ahead of the decision-making, however the leadership does not make it the priority over traditional practices.

In the end, the organizations that learn the fastest will win. If your organization can do most of its learning in decision-focused sprints that are days or a few weeks long during conceptual design, while your competition requires months to do the same learning in multiple cycles of detailed design, then you will soon be delivering results that they cannot touch.

Part II is a fictional story designed to allow you to experience such rapid focused learning on a complex problem (aircraft mission design) crossing multiple corporate boundaries. As you read, consider the competitive implications of organizations that can operate on a learning cadence measured in days competing with organizations operating on traditional learning cadences measured in months.

A STORY OF SET-BASED CAUSAL MAPPING ACCELERATING COLLABORATIVE LEARNING

II

Part II is a fictional story designed to let you experience a "greatest hits" of our many coaching sessions of many different teams across many different companies. It is our hope that this will give you sufficient experience with the real-world application of these techniques that you will be able to do the same for your own problems in your own organizations without needing a coach to guide you the first time through.

Chapter 7

Introduction to the Story

So now you have our "True North" statement acting as a compass pointing the way to establishing that "Success is Assured" prior to making key decisions. Further, you have seen the *Enablers* that make it possible for your development organizations to act on that pointer and establish that "Success is Assured" prior to making decisions. However, there are numerous co-dependent paradigm shifts in those Enablers, such as

- The shift from *point-based* to *Set-Based*
- The shift from *accelerating decision-making* to *delaying decision-making*
- The shift from *risk management* to *risk elimination*
- The shift from *schedule-based decision-making* to Decision-Based Scheduling
- The shift from *knowledge management* to focusing on and changing how you use knowledge in the first place in order to establish *systemic knowledge reuse*

It is very difficult to achieve any one of those without also achieving the others. Thus, you cannot work on one at a time; you need to make those paradigm shifts together.

By the nature of paradigm shifts, it is hard to imagine your future in the new paradigm while your thinking is still in the prior paradigm; you can only interpret what you are hearing in terms of what you know. So, until you have experienced at least some of that in your own work (some of you may have some such experiences to draw on), it will be difficult (if not impossible) to foresee that future state.

Our normal way to help organizations through that gulf is to come in and coach them through the application of the new way on their own problems or projects. In that, we are able to give the team a shared experience that they can then build on, establishing a deep understanding of "True North" and the various paradigm shifts they are making from their prior practices, which would otherwise remain far too easy to fall back into (the big hurdle in changing paradigms).

In this chapter, we meld together the big "aha" moments from hundreds of such coaching sessions into one combined fictional story designed to let you experience putting together much of what we have presented in the preceding chapters. To make it concrete, we are presenting a specific example; however, we have tried to broaden it enough that it can be successfully mapped to most any complex decision-making situation. While certainly not the same as experiencing it on your own problems with your own team, we do believe this chapter will allow you to "experience" it adequately and has the advantage that it is a "greatest hits" of such experiences.

In the end, great product development is about the effective collaboration of people to solve complex problems; thus, this story necessarily shows a collaboration. And to allow you to see the "worst case" scenario for collaboration, we show the collaboration across four different corporate boundaries and even more expertise boundaries. We will do so by continuing the story begun by Michael in his first two books, *Product Development for the Lean Enterprise* and *Ready, Set, Dominate*. However, there is no need to have read these books; the next few paragraphs will give all you need to know as background.

Michael's first book, *Product Development for the Lean Enterprise*, was published in 2003 and was the first published book describing the Toyota Product Development System. The book was about a failing infrared technology company, Infrared Technologies Corporation (IRT), that learned about the Toyota system and began to copy the practices of Toyota. The book registered with experienced product developers, as they could see their own organizations in the practices of IRT and contrast those practices with Toyota's. Based on our experience in working with dozens of organizations since then, those practices of IRT still represent the practices in most major development organizations today.

The second book, *Ready, Set, Dominate*, was published five years later in 2008 and reflected the results from the journeys of those organizations with whom we were working — some that were quite successful and some that never quite completed their transformation. The book introduced many important concepts such as the Knowledge Value Stream, LAMDA, and K-Briefs for problem-solving. Based on the common characteristics of successful companies, the book concluded with the following critical success factors:

1. Adopt a simple visual model of the transformation.
2. A learning process in place with a skilled workforce.
3. Functional subsystems are defined with clear responsibility for knowledge growth and quality.
4. An Entrepreneurial System Designer must be identified for every project.
5. Install a system for generalizing, visualizing, managing, and reusing Set-Based knowledge.
6. Top-level management must drive the transformation.

As before, we have continued to work with and learn from companies in the years since that book was published. We have found that while these critical success factors are still valid, they mostly define "what" needs to happen and not "how" to do it. In particular, how do you install an effective learning process that can rapidly identify and resolve all the critical knowledge gaps and trade-offs How can you develop a workforce with the right capabilities for execution? How do you build and use a knowledge management system that both reuses and continually innovates for the future? How does an Entrepreneurial System Designer optimize all the trade-offs on a schedule for market leadership? How does top-level management know if it is driving the transformation in the right direction? That is what this book is about: making clear *how* you do those things.

Introducing IRT

One of the common questions that we get is, "What happens when our organization adopts these practices, but the other organizations around us have not?" By continuing the story of IRT, we will be able to answer that by way of illustration.

In our continuing story:

- IRT has already incorporated "True North" into their leadership practices and has been establishing that "Success is Assured" prior to making key decisions.
- They consistently exhibit LAMDA practices around the visual models organized into K-Briefs for facilitating the LAMDA Discussions.
- K-Briefs have largely supplanted PowerPoint slides for engineering meetings, discussions, and presentations.
- Causal Mapping has become the key tool for root cause analysis, for identifying and resolving knowledge gaps (Decision-Focused Learning), and for trade-off analysis, enabling Set-Based convergent consensus (knowledge-based) decision-making.
- They have made the shift to Decision-Based Scheduling via Integrating Events.

In other words, they have put in place all of the Enablers discussed in Chapters 3 and 4.

They realigned their organization as shown in Figure 7.1 to support their behaviors and practices. A key role to making it all work was standing up the Functional Excellence organization under David Zeller, a well-respected technical leader who authored several of IRT's important patents. All of the development engineers report to functional leaders based on their technical expertise. These functional leaders are not only responsible for the technical execution of their functions but also for standardizing the resulting knowledge and continually innovating the knowledge. The functional leaders report directly to David.

The system designers and program leaders report into Jack Hawkins on the program side. Each of the business markets has a business leader responsible for

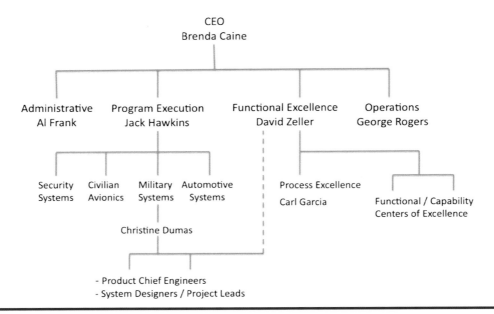

Figure 7.1   IRT's organizational structure.

project success in meeting profitability goals and customer interests. For example, Christine Dumas leads the Military Market Segment. Each specific product has either a systems architect and program manager team or a single Project Chief Engineer who handles both roles of systems design and administrative leadership of the program.

As shown in Figure 7.1, these project leaders report on a dashed line to David Zeller. This is not a reporting relationship but a responsibility relationship. Brenda Caine, the CEO, has given David the primary responsibility for achieving first-pass quality (i.e., establishing "Success is Assured" prior to making key decisions); as such, David chairs all of the gate reviews. In the past, these were largely task and schedule reviews in typical phase gate fashion; now, while the major review points are still relatively the same in order to align with customer milestones, the reviews are focused primarily on consensus decision-making based on validated knowledge. In addition to the primary gate reviews, additional Integrating Events are scheduled as needed for knowledge reviews and key convergence decisions.

Through these practices, IRT has established itself as a top-tier supplier of highly innovative technical solutions in all markets, and they deliver both on schedule and within budget. However, the transition hasn't always been smooth and is incomplete. Early on, the demand from the top for "Success is Assured" didn't match the engineering capabilities to do so. While they insisted all knowledge gaps be resolved before detailed design, unforeseen knowledge gaps (surprises) kept showing up late. The root cause was found to be that the engineering teams had insufficient means to recognize the knowledge gaps and trade-offs early in the process; only after geometry was developed could all the knowledge gaps be visualized. An important remedy was to introduce Causal Mapping as the means to identify the decision interrelationships and trade-offs earlier. While Causal Mapping provided great promise for early identification of knowledge gaps and trade-offs, there remained a problem: how do you integrate this capability into a systems engineering process that is already overburdened and focused on requirements flowdown — largely driven by customer requirements that are seemingly set in stone?

Carl Garcia, a top-level systems engineer and an original member of the team that studied the Toyota process, was added to David's team to resolve this and any other process or methodology problems. Carl developed and is implementing a three-pronged plan:

1. Build a team of Causal Mapping experts that can teach and coach teams in the methodology of using Causal Mapping techniques to identify decision interactions and trade-offs, and to innovate to achieve an optimized solution set.
2. Use this Causal Mapping team, officially known as the "Success is Assured" Mentors (or SAMs) to intervene with existing programs to ensure critical knowledge gaps and trade-offs are identified prior to approval of the system architecture leading to detailed design. Since David chaired the major program reviews, he could decide when the SAMs would intervene.
3. Use the LAMDA-based K-Briefs as the primary knowledge development and communication tool, not only internally but also with technology customers (e.g., DARPA). The goal was both to separate IRT from the competition by providing much richer technology transfer and to build deeper understanding of the desired trade-offs of their customers to better direct technology development for future applications.

So this is where IRT was at the start of our story. The five expert SAMs have been in place for a couple of years and have made a dramatic impact on back-end firefighting for ongoing programs. Program leaders take pride in not needing a SAM intervention, so they have been rapidly incorporating Causal Mapping as an ongoing tool. In addition, the Causal Mapping has been very successful in winning contracts with DARPA, utilizing the various visual models to clearly demonstrate what knowledge must be developed to establish that "Success is Assured," thereby more than compensating the time, cost, and effort involved.

However, the bigger profits would not come until David and Carl could start establishing "Success is Assured" from start to finish on their major production contracts, where these benefits would extend into the manufacturing, delivery, and service systems developed around the core product. An impediment to that was that IRT is primarily a supplier to larger contractors who are responsible for the complete system. As such, the lead contractor assumes the right to dictate schedules and hard specifications that force point-based behaviors, including unavoidable knowledge gaps and inevitable rework and firefighting. IRT stood out in that environment, and those contractors had come to prefer working with IRT and its superior knowledge, responsiveness, and flexibility. But IRT continued to be frustrated by the unrealized potential. Changing the mindset at those contractors had proven to be very difficult and frustrating, as well as severely impacting potential profitability.

So that is where we pick the story back up at IRT...

Note: if you lose track of who is who in the following story, we have a list of all the speaking characters in Appendix I.

Chapter 8

Monday Morning, Infrared Technologies Corporation

It was 7:45 a.m. as Alex Taylor was arriving in the front parking lot. While his career direction had been quite frustrating to him over the last year, today was even more frustrating. As a member of Carl Garcia's "Success is Assured" Mentor (SAM) team, Alex had been using Causal Mapping to analyze a team's design of a new radar guidance and auto-braking system, and to gain an understanding of their knowledge gaps and their plan to resolve them effectively. Although technically challenging and providing him very broad exposure to lots of different design challenges, Alex really wanted to be able to lead a design from start to finish. He enjoyed getting projects back on track but rarely got to enjoy being on track. Unfortunately, Alex was really good at doing the Causal Mapping and coaching the teams. While he understood the importance of his role, it didn't lessen his frustration. But sometimes, like that morning, there was additional frustration...

As he entered his office space, one of the other SAMs, Brett Lee, commented, "Good Morning, Alex. You look like someone stole your breakfast; you Okay?"

"Sorry Brett, I've never been good at hiding my feelings. I am presenting to the REC team the Causal Maps and Trade-Off Charts that I worked up for them last week; it makes clear that they overlooked some fundamental knowledge gaps that they really should have seen, and would have seen weeks ago if they would have done the Causal Mapping that David requested and they argued was unnecessary. Furthermore, I see no way to sugarcoat it to David when he sees the K-Brief. These guys are my friends; I want to work with them in the future. I have been dreading this meeting all weekend."

"Geez," Brett shrugged, "give yourself a break, it would be a helluva lot worse if those knowledge gaps had shown up later, after they were in testing; they'd be working long hours to make up for that. Has Carl given any indication when you will be transferred back to the projects?"

"No, not a word since we last talked about a month ago. I don't think David wants to change anything right now since we are really catching a lot of knowledge gaps that would have otherwise escaped. And the projects are starting to ramp up on their own, but they still need a bit of coaching to get going."

About then, the phone interrupted the discussion. Looking at the phone's display, Alex said, "Speaking of the devil: it's Carl on the line."

"Oh, maybe he was listening."

Alex lifted the handset. "Good Morning Carl, you're calling early — makes me worry a bit."

"Alex, you worry too much," Carl laughed. "Can you come to a meeting tomorrow at 8 a.m. in Brenda's conference room?"

"Yes, certainly; should I prepare something?"

"No, we have a new opportunity to discuss. Do you have a few minutes this afternoon and I will brief you?"

"No, not today. I am working with the REC program all afternoon."

"Yeah, I don't want to disrupt that," Carl agreed. "I will try and catch you before the meeting tomorrow; you might come a little early."

As Alex hung up, he felt both a little hopeful and a little worried, as he was unsure what to expect; it might be important if it's scheduled in Brenda's conference room.

Chapter 9

Tuesday Morning, Infrared Technologies Corporation

Alex arrived at his office at about 7:30 a.m. He quickly checked his email, grabbed his computer, and headed to the meeting, hoping to catch Carl before the meeting started. As he walked to the executive area, he was much less frustrated than the day before — the meeting with the REC team ended up going really well — yet he still had some apprehension as he approached Brenda's conference room.

No sign of Carl as he entered the conference room. He really liked this room: mahogany table with all the computer connections and mics one can imagine, dual screens, video conference cameras — the works. The walls were covered with pictures of IRT products. He took a place at the back of the room, set his computer there, and headed out to grab some coffee in the little side room.

As he re-entered the conference room, David Zeller followed him in and said, "Hey Alex, you got here early. How've you been?"

"I'm doing great. I arrived early to hopefully see Carl for a briefing on this meeting; all I was told was to show up."

David laughed, "Yeah, we have a bad habit of doing that to folks. This is a pretty exciting opportunity. How did the meeting go with the REC team yesterday?"

"Great. The whole team was there; Sue Giles from Structures attended for the first time and really added a lot. They actually took over from me, driving the Causal Mapping themselves, adding more decisions and relations, and innovated some really great alternatives. They now have a good plan to resolve the gaps and converge. I basically just asked questions to make sure they didn't run too quickly past some of the potential knowledge gaps they were making visible."

"That's excellent to hear. That's been our goal with your team — to be a catalyst for the organic growth of the methodologies — so that is encouraging."

About that time, Jack Hawkins, David's counterpart over Project Execution, and Christine Dumas, over Military Products under Brian, came in together.

David, Brian, and Christine began talking management talk. That gave Alex an opportunity to refill his coffee.

Outside the room, Carl was just coming in. "Sorry I'm a little late — got held up at home."

"No problem," Alex shrugged. "Is there anything I need to know?"

About that time, Joe Rivera walked by them and into the conference room. Joe is widely acclaimed as the top systems designer/chief engineer/program manager at IRT. He is one of only two that has the title of Program Chief Engineer, where he is both the systems designer and the administrative leader, similar to the role of the Toyota Chief Engineer.

"Is Joe going to be involved in whatever this is?" Alex asked.

"Yes, he is going to be the Program Chief Engineer. You are going to be his assistant."

Both excited and concerned, Alex asked: "Does Joe know this? I think he hates me. Do you remember how angry he was with me when David had me lead a Causal Mapping session on his project a while back?"

"Alex, you worry too much. He wasn't mad at you; he was mad at himself for not seeing what you pointed out. He actually asked for you to be his assistant — insisted, actually. We need to get into the room."

As they entered the conference room, Joe had taken the spot next to Alex's computer. As Alex sat down, Joe simply said, "Good morning, Alex, good to see you again."

David opened the meeting stating there would be no slides and it would be short, for which he received grateful responses.

"We have an opportunity to achieve what has always been the larger goal with our 'Success is Assured' efforts: to execute and demonstrate the power of this thinking from the very start of a complex product development project. As you all know, we have been very successful using the methods on technical study projects and for eliminating loopbacks on existing projects, but we haven't yet had the opportunity on a project from start to finish.

"We all know our basic problem: we are a supplier. Our contracting customers still work in detailed spec flowdown, which drives point-based non-optimized solutions; as a result, while the benefits have been great, there's always so much more we can clearly see that is being left unattained."

Sipping on his coffee, David continued: "DARPA is actually promoting this opportunity, as they have seen the results from our Set-Based interactions with them around Causal Maps and K-Briefs. They have encouraged the Navy to work with us in advance of setting specs to get visibility to the larger design space. The Navy is really excited about several different mission profiles based on our Hawkeye Multi-Tracker device developed last year under a DARPA program. They want to be able to track multiple targets over a broad area for an extended period for search and rescue, extended reconnaissance, and other missions where they want to get in and out without detection. For that, they want to put up numerous trackers on drones to cover different size areas, with a certain level of redundancy.

"While the potential device sales are obviously great for us, the potential opportunity here is much greater. Given there will be major trade-offs to be made that cross the aircraft and our device, the Set-Based knowledge we have built can be leveraged to allow us to drive the development of the specs from the very beginning of the project, before it is even under contract."

Jack jumped in, "David, I understand the potential and you know I support it since I am allowing Joe to be the Chief Engineer, but do you have a real plan on how to actually make it play out that way?"

"Well, kinda. We believe that if we clearly lay out visually the key trade-off decisions that they need to make and use our Set-Based knowledge to show them the sensitivities and the bounds on the design space, then they will realize that they need to keep us involved throughout that complex decision-making. Further, we know they would love to establish that 'Success is Assured,' they just don't think it's possible to do so; rather, they just have to manage the risks. But if we can demonstrate our ability to help them actually prove that 'Success is Assured,' then perhaps we can get this project where we want it from the start.

"So, the key to our success is keeping the focus on the knowledge gaps that need to be closed to make the right decisions on the design of the larger system, which are ultimately gated by our knowledge. The longer we can keep the program focused on closing knowledge gaps prior to developing specs and contracts, the longer we can delay a fallback into traditional practices, at which point they likely move the aircraft company into the prime contractor role and we end up back on the tail of the dog."

Jack chimed in: "Where we always seem to end up!"

David nodded. "That's where Joe and Alex's roles become critical. It is not good enough for *us* to see the knowledge gaps; we have to make the gaps clearly visible to the Navy. We want them uncomfortable moving forward without first closing the key knowledge gaps. We need to make as many of the trade-offs visible as possible, such that they understand the impacts those trade-offs will have on them. We want to help them design the path to 'Success is Assured,' such that we can put ourselves in the middle of that path. Joe, do you have anything to add?"

Joe slowly stood up and walked to the front of the table. Where David tends to speak with passionate intensity, Joe tends to speak with a more contemplative calm, though perhaps with even more intensity. "Not much, just this: We are all familiar with the current system made up of customers passing unrealistic specs and schedules to the prime contractors, and then those being broken down into unworkable subsystem specs and schedules for the suppliers, and all the resulting firefighting that comes from that. That system is not delivering profits to anyone in that chain and is consistently not delivering what the end customer really wants. That has never been made so clear as it was on the last project I worked on with Alex; that Causal Map made the stupidity so obvious to everyone on our team that it was almost intolerable. But we had no access to the end customer, and our customer was too swamped to consider anything different. We have a rare opportunity here to involve the end customer directly; we need to pull out all the stops in developing that visibility. We need to make sure falling back into traditional practices is completely intolerable to the Navy, such that when they get a more optimal product on time and on budget they will be fully motivated to make that change permanent."

David concluded the meeting by stating that anyone Joe and Alex needed from his organization would be made available. Everyone seemed ready to go, and they closed the meeting.

Joe turned to Alex. "Alex, we're meeting with the Navy at 9 a.m. tomorrow; they asked us to block off all day but didn't give any more details on the agenda. They are in town this week with plans to visit UA to discuss the unmanned air-craft part of this on Wednesday and wanted us to have time available Thursday and/or Friday as needed to answer UA's questions. So, we need to turn that around tomorrow as best we can. As prep, can we meet at 1:30 this afternoon to go through the Hawkeye Multi-Tracker Causal Map that you developed with that team?"

"Sure. I'll need to catch up with where they took it. I haven't seen the knowledge that was developed to close some of the key gaps. But I was in the knowledge review prior to the delivery to DARPA, so I know they finished it out and that all the key trade-offs were made visible. The full design space was mapped out."

In that afternoon meeting, Alex quickly brought Joe up to speed and they strategized on the key trade-offs the Navy would likely be running into. But the Navy was driving the meeting, so Joe's primary message to Alex was this:

"Alex, don't hesitate to do what you always do in these meetings internally. Make visible what they are asking for and make visible the knowledge gaps — what's needed to establish that 'Success is Assured' for their goals. Don't hesitate to jump in and focus the conversation around the visual models, particularly the Causal Map that I know you'll be drawing as soon as they start talking. We obvi-ously do not want to offend them, but let me worry about managing that. This may be our one shot at getting this on the right track; I'd rather go big and have to make amends than go soft and get left in the traditional rut. The Navy should feel like working with us is completely different from working with any of their other contractors."

Alex responded with a relaxed "Okay, this should be fun," hiding both his excitement at the potential and his trepidation at not knowing how it would play out and having zero experience in front of a customer. But in the end he knew the power of visual knowledge to drive efficient collaborative discussions, even in the face of great incoming contention, so he was not too uncomfortable with his role.

Chapter 10

Wednesday Morning, Infrared Technologies Corporation

Not wanting to take on entertaining their Navy customers ahead of the meeting, Alex decided it would be wise to arrive just before 9 a.m. As he entered the conference room, computer in one hand and coffee in the other (i.e., ready to go), he identified three visitors from the Navy; the whole group was standing at the front of the room. The largest, an imposing man with a deep boisterous voice, was wrapping up a story he was telling to Joe and David. Alex spotted Joe's computer and set himself up in the chair left of Joe.

As the story ends with some laughter, the storyteller switched gears: "It's nine, can we get started? We're on a tight schedule this week, so we would like to make sure we get all we need from you today."

As everyone moved to their chairs, Joe replied, "We are ready to go. You have the lead."

"I am Commander Nathan Harris. I am responsible for mission support systems, particularly the larger surveillance and coordination needs. To my right is Lieutenant Commander Mark Jackson. He is a surveillance systems expert and will be the subject matter expert in this acquisition. To my left is Commander Laura Ramirez. She is the acquisitions lead; she will handle all the legal and contractual negotiations. Mark and I are focused on figuring out what it is we need, what it is we should be asking for; Laura is responsible for all the rest involved in acquiring that.

"As you know, we called this meeting because we are very interested in using the new Hawkeye IR tracking device that you developed under a DARPA contract. We want to use it for various situations: tracking enemy targets in battle situations, tracking targets for search and rescue missions, and tracking targets for intelligence gathering. However, to cover different-sized target areas for different periods of time, we need to understand how many devices we will need and therefore how many unmanned aircraft we need."

Laura added in the real driver for why they had bothered to come and visit IRT: "Our understanding from DARPA is that you guys have a scalable solution and that you have a sophisticated model of how the solution will perform in different scenarios. We want to make sure we understand that, such that we can optimize the specs to fit our mission needs."

David, thrilled to hear that their intent was precisely what he wanted it to be, replied, "Yes, we have developed an extensive causal trade-off map on how it performs, allowing us to quickly make the design decisions required to meet different sets of customer interests."

"Great." Nathan then handed the lead to Mark: "So, Mark, you have the questions…"

"Yes. First, how large an area can one device cover?"

As IRT's technical lead, Joe replied, "That depends on a lot of things. First, how small a target do you need us to detect and from what distance?"

"Well, that was going to be my next question: what is the smallest target you can track from 40,000 ft versus 60,000 ft versus 80,000 ft?"

Joe tried to clarify that there's not a simple answer: "We can detect most any reasonably sized target of adequately different temperature from any range, given a big enough receiver and enough power. How big is the aircraft? How big of a diameter receiver can we put on it?"

Mark chuckled, "Yeah, that was going to be my next question for you: how big and how heavy are your devices? We need to know that to size the aircraft."

Nathan chimed in, "And the size of the aircraft is going to dictate how high we need to fly in order to avoid detection by enemy radar."

With a bit of a sigh and an odd look over at Laura, Mark added, "That's just one of the circular dependencies that makes it very difficult to figure out what we need to be putting in an RFP for the aircraft companies."

Laura, sounding a bit annoyed or disappointed, interjected, "We were hoping you could help us break out of that."

Mark, revealing himself to be the less optimistic of the three, followed up: "But it sounds like you are just adding some more circular dependencies. To do these sorts of analyses, we need to start with an airplane design. I suggest we assume the UA-243 aircraft; I know its specs. I can answer their questions based on the UA-243, and then we can see how the Hawkeye will perform on it. Given how far along that design is, it is going to be hard to beat."

Now openly irritated, Laura responded almost harshly, "Mark, we've discussed this. The optimization of this design will be critical; the last three projects in this area have all been terminated after much expense because they ended up being unattractive. I do not want us jumping to solutions."

Mark pushed back with what seemed to be the continuation of a longer-running argument between the two: "We can still optimize. But we have to start with a design that we can analyze; otherwise, we don't know what we're optimizing."

David, wanting to make sure the typical circularities didn't abort the discussion prematurely, interjected, "Well, I am confident we can help you work through the circular dependencies. However, we may not have all the knowledge needed; we may need to pull in some experts in other areas."

Joe added in his calming analytical voice, "Agreed. But first we need to get clarity on what you are trying to accomplish and what we know and don't know."

Joe was looking at Alex as he said that last phrase, giving Alex the confidence to make his standard entry into such technical discussions. "I've been taking notes in the form of a Causal Map; perhaps it will help." He pressed a couple buttons on the touchscreen mounted on the end of the high-tech conference table and his Causal Map (Figure 10.1) appeared on the two screens at the front of the room. "The shapes marked with a red target are the things you asked for; the rest are the things Joe asked you in order to answer those questions. Do I have this right so far?"

All eyes perused the map on the screen.

Nathan saw the different things that they had asked and the various connections but immediately thought of a number of others that should have also been connected but which they hadn't asked yet. But rather than answering Alex's question by volunteering those, he instead inquired as to where this was going: "So, if we get all of our needs laid out here and connected into your map, how is that going to get us answers to our questions?"

"Ah, good question," Alex replied. "As David mentioned earlier, we have an existing Causal Map that connects requirements such as these to our own internal design decisions that will allow us to compute what is possible — to compute the feasible design space. Let me drop in a cloud representing all that internal stuff that we need to decide to know the trade-offs between these things."

As Alex dropped that cloud into the Map and connected it up, he continued, "The Map represented by this cloud shape covers all the circular dependencies in our part of the system; what we want to do here is capture those circular dependencies in that larger system that you are trying to work out, so that

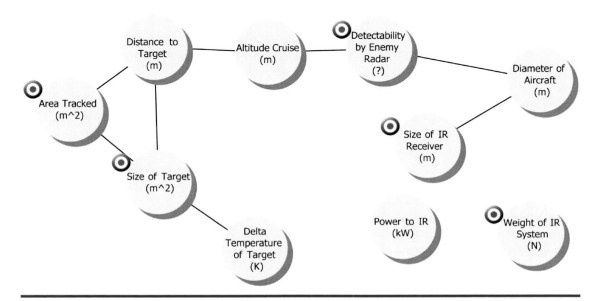

Figure 10.1   Alex's initial Causal Map of the Navy's problem.

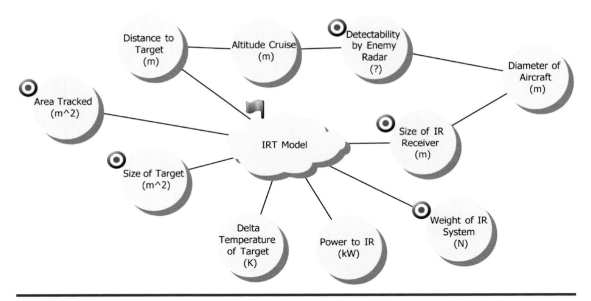

Figure 10.2 Adding in IRT's existing model as a cloud shape (details elsewhere).

we can help you work through those." (Figure 10.2.) (*Note:* we will place a flag on each shape that was added to the prior Map to make it easier to spot the changes as the Map evolves.)

> As you read the rest of this chapter, to get the full experience we are trying to share with you, you need to really read the Causal Maps in detail as if you want to understand, critique, and improve them. If you find yourself confused about what was added each step of the way, then we might suggest that you are not really reading the Map and critiquing it deeply but rather just looking at it as if it was a cartoon where the details don't matter. The Causal Map is all about making the details visible such that they can be challenged and corrected to establish "Success is Assured" prior to decision-making. To experience that, you can't gloss over those details.

Mark gave Laura a skeptical look that Alex was all too familiar with — a look that said, "This young kid has no idea how complicated our stuff is. This is never going to work." Laura's return look also seemed skeptical but possibly more open to seeing where this might go.

Nathan did not see either look as he was hyper-focused on the Map. "Why isn't 'Altitude Cruise' marked with a red target? We normally spec that, so wouldn't it be a customer interest?"

Alex explained his logic: "It sounded like the reason you cared about Altitude was the Detectability of the aircraft; if we can reduce that by shrinking the aircraft, then a lower altitude would be fine."

"True… but Altitude affects other things, such as the range/endurance of the aircraft."

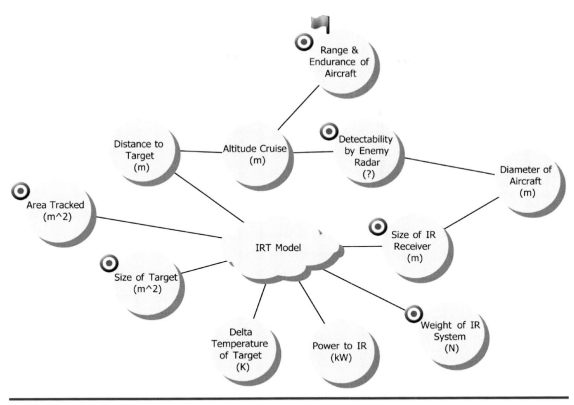

Figure 10.3   Adding 'Range & Endurance' as a key customer interest.

As Nathan said that, Alex added in an additional shape (Figure 10.3) for "Range & Endurance of Aircraft." "Okay, then I'd mark those other things with the red target."

Nathan, starting to get a feel for the Causal Mapping, pointed out some other potential flaws: "Well, the Range and Endurance will also be dependent on the Diameter of Aircraft and Weight of the IR system."

Nathan's focus on the map had pulled Mark in, "Indirectly... both of those will affect the Weight of the Aircraft which will affect the Range and Endurance."

"Gotcha," Alex replied, as he added the additional connections into the Map.

Given that, Mark began to question some of the target symbols that Alex had: "So, from that standpoint, although we are going to ask you for the weight of the IR system, we don't really care about that except for its impact on Range and Endurance."

"Ah, then you're right, it probably shouldn't be marked with a target, and similarly on the Size of the Receiver."

Continuing to walk around the map (Figure 10.4), Nathan asked, "What is the power source for 'Power to IR'?"

Joe replied, "Well, it could be the aircraft. We could power it with batteries, but those would be very heavy, so I suspect that would be a big enough drain on the aircraft that it would be better to just draw power off the aircraft."

"Well, that will directly impact 'Range and Endurance' as well then," Nathan added.

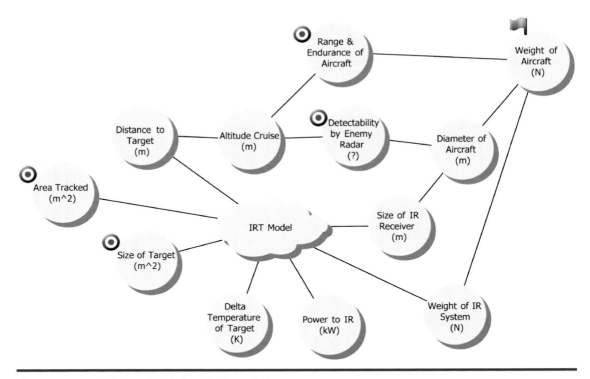

Figure 10.4   Adding 'Weight of Aircraft.'

"Okay, let me connect those." Alex rearranged the 'Power to IR' node to the top to make it easier to connect to 'Range and Endurance' (Figure 10.5).

Joe, seeing a missing connection, added, "At higher altitudes we'll have colder ambient temperature, which will help us keep it cool, reducing power consumption. So, 'Altitude' should also tie into the IRT Model through ambient temperature."

Mark, being a stickler for correctness, added, "Well, it could get hotter or colder as you rise; but yeah, that doesn't change the need for the connector in the Map."

"Okay, I have added that in." (Figure 10.6.)

Laura decided to check her understanding: "So, this map is connecting various attributes of the system design. 'Altitude Cruise' is how high we want to be flying at, which in your model is 'Distance to Target.' Your model of the Hawkeye device takes in that 'Distance to Target' plus 'Area Tracked,' 'Size of Target' to be tracked, the temperature delta of the target versus the environment, and 'Ambient Air Temperature,' and from that it gives us 'Size of IR Receiver,' 'Weight of IR System,' and 'Power to IR.'"

Alex cut in with a correction, "Well, it could give that, but it would be a set of possibilities. If you want 'Size of IR Receiver' to be smaller, we could possibly do it, but it would require more 'Power to IR.' Or vice versa, if you have an upper limit on 'Power to IR,' we can do that but we may need a larger 'Size of IR Receiver' or heavier 'Weight of IR System.' Or we could trade off against any of the other values as well. In other words, you can decide any seven of those eight decisions, and we'll tell you what the eighth needs to be."

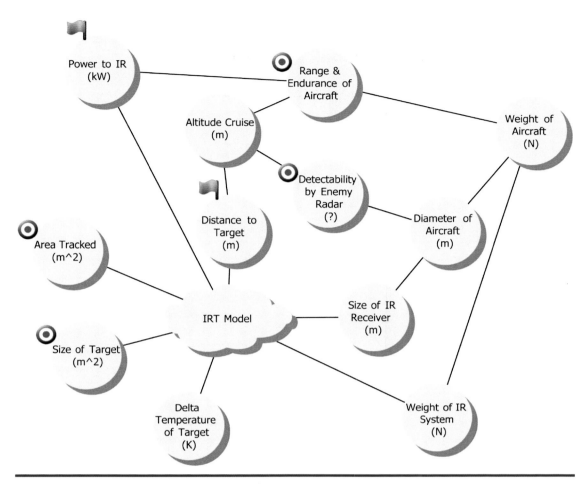

Figure 10.5   Connecting 'Power to IR' to 'Range & Endurance.'

"Okay," Laura continued hesitantly, "so 'Power to IR' is going to impact 'Range and Endurance' because it will be consuming power from the aircraft's engines. 'Range and Endurance' will also be impacted by 'Weight of Aircraft,' which in turn will be impacted directly by 'Weight of IR System' and indirectly by 'Diameter of Aircraft' driven by 'Size of IR Receiver,' since presumably a bigger-diameter aircraft will weigh more. Am I reading that right?"

Alex answered, "Yes, exactly."

To this, Laura asked, "Okay, nice picture, but what does this tell us?"

David explained, "It tells us *what* we need to know to answer your questions. Now we can go through this and determine what of that we *do* know and what we *don't* know. Then we can focus on closing those knowledge gaps so that we can answer your questions with confidence."

Joe elaborated, "For example, to get more 'Range and Endurance,' you might ask us to reduce 'Power to IR,' but in doing so we might need to increase 'Weight of IR System,' but that in turn would increase 'Weight of Aircraft,' which might end up reducing 'Range and Endurance' more than we saved with the reduced 'Power to IR.' This Map helps to make sure we understand all that will be impacted by our decisions."

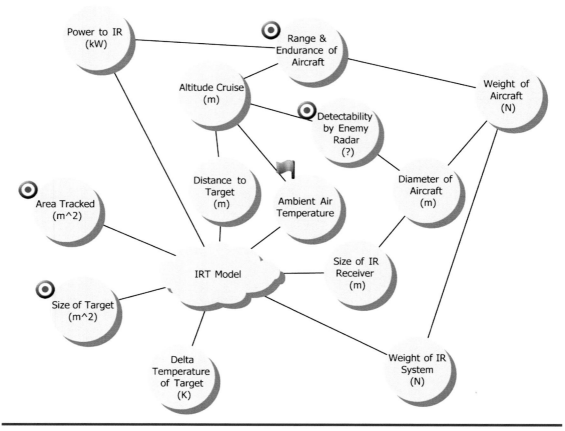

Figure 10.6   Connecting 'Ambient Air Temperature' to 'Cruise Altitude.'

Laura decided to give them the benefit of the doubt for now with a hesitant "Okay."

Alex returned the focus to fleshing out the Map: "Well, first we have a few holes in what we've captured here. What is the unit of measure of 'Detectability by Enemy Radar'?"

All eyes turned to Mark. "Hmmm… We would probably measure it as 'Radar Signal Level' of the aircraft on enemy radar. I'm not sure the unit of measure of that, but I know who would know."

"Okay, I'll leave that marked with a question mark; the 'T' means we think we know where to get this, but let's put who we think knows that in the notes on this shape so that we can actively close that knowledge gap."

"You can put 'Ask Lisa' on that," Mark added.

"What's the unit of measure for 'Range and Endurance of Aircraft'?"

Mark pondered a moment. "Well, those are two separate things: Range is measured in kilometers or nautical miles; Endurance is measured in minutes."

Alex split that Decision shape into two and added the units to each. "And what is a reasonable range of values for those? I guess that's a question for the Navy, since those are customer interests. For the missions that you are considering, what are the smallest and largest range and endurance values you might need?"

"Let me think," Nathan paused. "Do you want typical values or the most extreme values?"

"The most extreme. When working Set-Based, trying to establish that 'Success is Assured,' we want to make sure our design can cover all missions that you are interested in covering. On the other hand, we don't want to waste time or money trying to make it handle missions that you would never be interested in. So, we're trying to eliminate uninteresting parts of the design space."

"Okay, that makes sense."

Alex continued, "Also, you'd be surprised how often making visible the ranges of values in different people's heads results in finding key misunderstandings or exposes key knowledge gaps that otherwise would have gone unnoticed."

"I think we'd be safe setting the maximum Range at 3100 nautical miles. The minimum would never be zero, but it could be pretty small. Let's call it 100 nautical miles. On the Endurance, let's use five hours (or 300 minutes) as the maximum. Not sure on the minimum."

Mark jumped in: "Well, given we are modeling the possibility of being detected or attacked and needing to turn and run, I would say that could be as low as 0, if we were detected immediately upon arrival."

Alex responded, "Yeah, that makes sense, so I have captured that."

Throughout most of the rest of this story, we will skip the discussion and capture of the ranges of values for each Decision, given that we don't want to bore you with such details and this is not what you are trying to design. However, when doing your own design work, don't forget to ask those key questions. Not only do they often uncover misunderstandings between different people and groups, they also help ensure that people are considering the right ranges of values.

For example, when someone hears a larger range than they are expecting, they may end up pointing out that if you want to go that high, that far, that fast, that hot, and so on, you cannot make the simplifying assumption you are making. For example, later in this model, when we get to the desired cruise velocity, if the maximum value exceeds the speed of sound, a great many of the equations would need to change.

Conversely, if someone hears a smaller range of values than they were expecting, they may volunteer simplifying assumptions that can be made. For example, when hearing that the maximum velocity is well below the speed of sound, they may suggest that some limit on the Fineness Ratio won't really have much of an impact, so perhaps it can be dropped from the Map entirely.

Looking at what Alex had captured, Nathan suggested, "Range and Endurance aren't independently driven by the same inputs; they are actually direct trade-offs. You can always reduce 'Range' to leave you more fuel for 'Endurance' or vice versa. So, I think you should add a line directly between them."

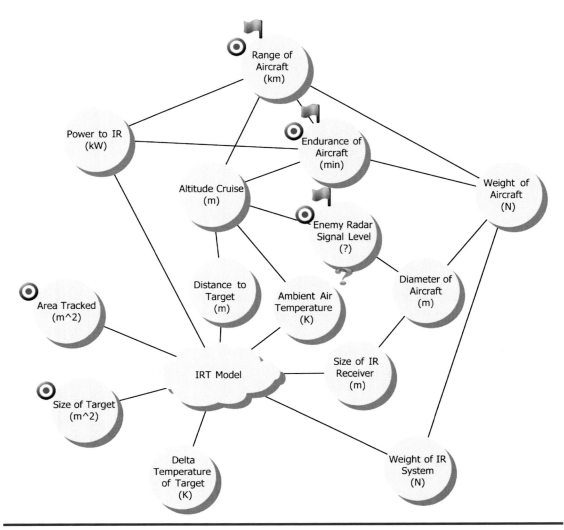

Figure 10.7   Splitting out 'Range' and 'Endurance,' adding in 'Enemy Radar Signal Level.'

Alex added the connection (Figure 10.7). "Okay, got that. The relations between these shapes on the bottom, connected by the 'IRT Model' cloud, are all known by us; we have a model for that. So, which of these other relations between the shapes do we know? Or know where to get the info? Starting from the lower right and working our way up, 'Weight of IR System' is obviously a direct add to the overall aircraft weight. Similarly, 'Diameter of Aircraft' will need to be bigger than 'Size of IR Receiver,' but how much bigger is not clear. I assume the aircraft guys could tell us that?"

Nathan replied, "Yes, and you'll need an aircraft company to translate that additional diameter into additional aircraft weight. They might also be able to translate that into additional 'Enemy Radar Signal Level,' but we might need to talk to a Radar company to get that. Oh, and the added payload weight for your IR system won't be a simple add: you'll need more wing and more structure to carry that weight."

Mark, always ready with more details, added, "The effect of 'Altitude on Ambient Air Temperature' can be pulled from the standard atmospheric tables.

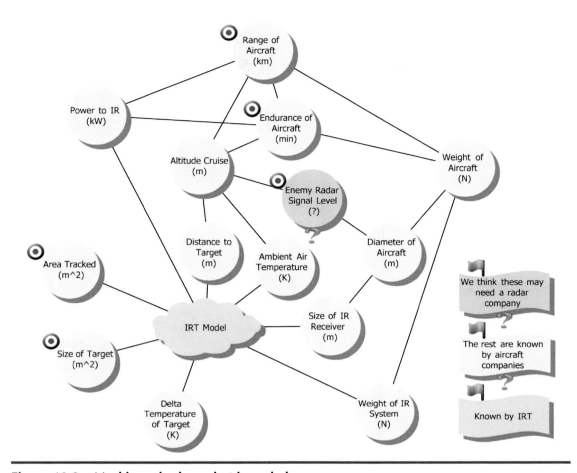

Figure 10.8   Marking who has what knowledge.

And 'Distance to Target' is pretty much identical to 'Altitude.' But the effect of 'Altitude' on 'Range' and 'Endurance' is complicated; we'll definitely need to work with an aircraft company on that."

With a bit of a squint at the evolving Map (Figure 10.8), Mark pointed out, "Well, really, the effects of added weight, altitude, and power draw won't have independent effects on 'Range' and 'Endurance' as its drawn; that's one big expensive analysis that the aircraft company is going to have to do by designing and analyzing and optimizing the aircraft. That doesn't seem to be reflected properly in this map."

Alex pondered for a moment, "Hmmm… You're right… I was trying to keep the Map simpler, but sometimes more shapes makes the Map much clearer; we really need to start capturing some of what you just said into Relation shapes. First, you said that we have a multiplier between the weights; we'll need the aircraft companies to tell us that." Alex added to the Map as he continued. "There is a similar adder between the diameters… and some more complex translation of the diameter into weight. And then the weights and power and altitude all go into some more complex model to get 'Range' and 'Endurance,' which can then be traded off between each other based on where we spend the fuel."

Mark asked, "Should you also add a note on using the standard atmosphere tables to translate altitude to temperature?"

"Oh, yes." Alex added that relation. (See Figure 10.9 for the added relations, marked with blue flags.)

Laura was starting to see how this would make their path to getting the answers she was looking for much clearer. "Okay, that clarifies the knowledge we need. I think we can get some of that for you quickly. But at least that top cloud and probably the other one are a chicken-and-egg problem for us. The aircraft company will need to design and optimize an aircraft — weeks or months of analysis — but to do that, they'll need the requirements from us; but we're trying to gather this knowledge to help *you* help *us* figure out those requirements."

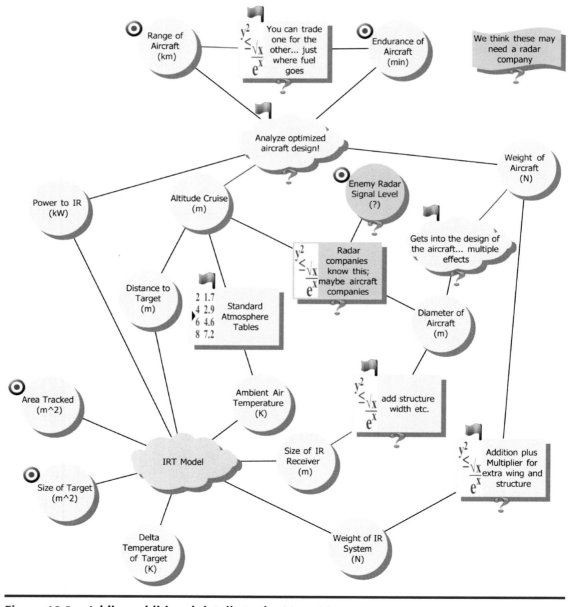

Figure 10.9   Adding additional details to the Navy Map.

David had heard that objection countless times on this journey and in this case saw it as an opening for advancing their larger strategy. "Yeah, we used to tell our customers similar things: that we can't know how an IR design will perform until we design it, build it, and test it; that it's just too darn complex; that dealing with the noise in different mediums is not easily modeled; you need that CAD design. We eventually learned to think in terms of worst case, design limits, and sets of possibilities. We have gotten pretty good at coaching our suppliers through these analyses. They have the knowledge we need, and we help them visualize and organize it, just like we're doing here. We haven't yet done that with one of our aircraft company customers, but we're willing to do that — as long as they are open to thinking about this a little differently. And hopefully we'll break down these clouds into knowledge gaps that they know how to close — no 'problematic' knowledge gaps."

"We can pull in an aircraft company." Laura paused as she considered how this would likely play out. "But it'll be tough to get much time from them if we're not paying them. And we can't just hire one; we would need to go through a proper RFP and selection process, and for that we'd need specs. Chicken-and-egg again."

As Laura was saying that, Alex marked those two clouds with "Problem" flags to indicate there might be some problems to be solved in getting that knowledge.

David, seemingly impossible to discourage, probed for the right "carrot" to use. "What if we can convince them that the knowledge would be reusable on future projects and would make them far more efficient and effective in product development? Twofold improvements in productivity or more. Would that entice them to invest a little time?"

Laura shrugged as she looked over at Nathan and suggested, "How about UA? They've been promoting their continuous improvement efforts as a reason for us to go with them. And we're already scheduled to meet with them tomorrow. Perhaps we'll challenge them to put their words into action."

Nathan answered, "Yeah, they are a sharp bunch; they know their stuff. I think they might actually enjoy this process; I know I have so far."

Laura sighed at the action item she just took on. "Alright, I'll contact UA this afternoon and try to make sure the right people will be in the room tomorrow. Anything else at this point?"

All responded with a collective shrug.

With a smile, Nathan said, "And I was expecting this to take all day. We didn't even make it to the morning break."

The visual model (shown in Figure 10.10) not only helped the team quickly share what they knew, it also made it clear when the meeting was done: when the knowledge gaps are all things that you can't fill in with the people in the room, it's time to make plans on how to fill them and get on with it — very different from traditional meetings where teams will go around

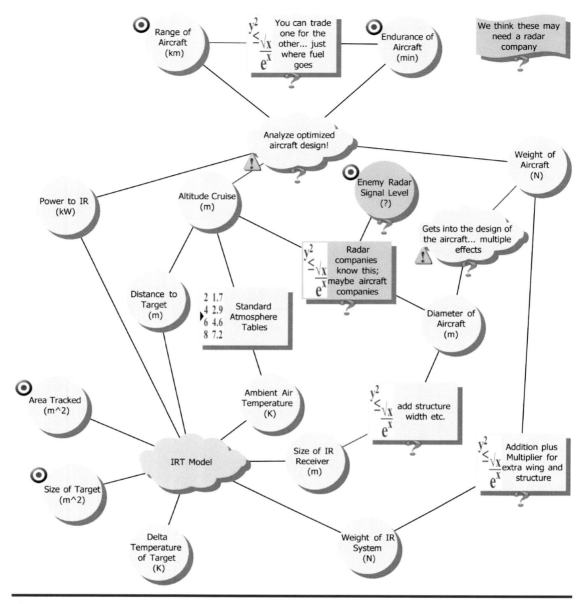

Figure 10.10 Navy Map with knowledge gaps marked.

and around fretting about all the complexities and cross-dependencies but never really sorting them out, never creating any clarity on what is known and what is not, and never making any concrete plans on how to efficiently move forward. Those meetings tend to end when they run out of time (encouraging most to try to keep them short, but the people needing the answers trying to make them long in hope that they'll get to an answer before it ends).

For those of you who take pride in your Sherlock Holmes-like skills in solving the mystery or twist of the novel or motion picture long before it is revealed by the story's characters, we challenge you to do the same here. What are those key facts that are revealed in the story but do not get incorporated into the characters' or readers' causal logic until much later? For those of you who are experienced in Causal Mapping, you can use the following chapters to hone your skills. What does Alex fail to add to the growing Causal Map until someone else points it out (again) much later in the story?

Chapter 11

Thursday Morning, Unmanned Aircraft, Inc.

The IRT team (Joe and Alex; David had other meetings to attend) decided to carpool from IRT to UA's facility so that they could get in sync on the morning strategy. So, as they walked into the very modern glass lobby together, Joe commented, "Wow, you can feel the Silicon Valley influence here — very high tech! Nice."

The Navy team (Nathan, Mark, and Laura) were already in the process of signing in at the security desk. As the IRT team joined them, Nathan asked, "Alex, were you able to print up copies of the Causal Map as we discussed?"

"Yes, I printed extra; often we end up drawing on them and then end up wanting fresh copies to take a different direction or dig into a different area."

Nathan nodded, "Perfect; in discussing the best way to engage with UA on this, we think it best to just start with the Causal Map and make it clear what we need from them."

Their UA host, Tim Lewis, then came through the fancy powered security gates, welcoming Laura with catch-up questions regarding some past conversation. That quickly transitioned into brief introductions of the other five at the security desk.

"We'll be meeting today in our main conference room right off this lobby. Teresa and Scott are already there." Tim gave the standard orientation to where the facilities were, including a pointer to the luxury single-cup coffee and espresso machine — which of course the IRT team happily visited on the way into the conference room.

Nathan opened the meeting with the same introductions as the day before. Joe introduced himself and Alex. Tim followed suit: "I am one of the program managers here at UA. Based on the expertise that Laura asked for, I asked Teresa Hernandez to join us as she is one of our most senior Systems Engineers; if she doesn't know the answers to your questions, she will likely know who does. Scott Park is my go-to guy in the Testing and Analysis organization. Whenever I want clarity on what data we have or don't have, Scott is my guy."

Nathan continued with largely the same opening as the day before, and then added a summary of the prior day's meeting: "So, we started asking these IRT folks the many questions we needed answers to so that we could develop specs to bring to you today. What we ended up getting was visibility of a lot more questions that we hadn't known to ask."

At this point, Alex was walking around the table handing each a copy of the Causal Maps he had printed.

"Alex is handing you each a copy of what we worked out yesterday with the help of the IRT folks. We want to continue fleshing this out with your help. In particular, each shape with a question mark (?) on it is something where we needed more expertise for clarification. But the key areas we need your help on are the two cloud shapes, where there's probably a lot more details needed. So, to start, Mark will lead you through this Map. And then he'll run down our list of questions and see how much you can answer for us."

Without looking up from the Map in front of her (Figure 11.1), Teresa interjected with a hint of reluctance, "We're happy to help, but to even semi-accurately compute 'Range' and 'Endurance,' we'll need to draw the airplane… and we'll need to know a lot more than what you're showing here to properly size that airplane."

Alex, accustomed to just such a response, was ready with an answer: "We don't need to know the 'Range' or 'Endurance' with much accuracy; we just need to know they will reliably be above the required levels."

"Okay." Teresa paused a moment digesting that thought. "That sounds helpful in theory, but I'm not sure how we practically do our analyses any differently."

"I understand, we hope to help with that part," Alex said in an assuring voice. "Perhaps the best way to start is to ask you what more you would need to know to be able to compute 'Range' and 'Endurance.' In other words, what belongs where that cloud is that is connected to 'Range' and 'Endurance'?"

Scott volunteered, "Well, we can use the Breguet Range Equations to get 'Range' or 'Endurance.' To feed those, you need the starting weight, the ending weight, the specific fuel consumption, the lift-to-drag ratio, and the velocity you want to cruise at."

Teresa added, "That's a *very* rough computation if you do it once for the whole flight. More properly, we break the mission into its normal segments: takeoff, climb, cruise out, loiter at the target, combat, cruise back, loiter to land, descent, and landing." Teresa simulated the flight of the aircraft with her hand as she said that. "The computations for each of those is somewhat different and much more accurate."

Scott acknowledged that complexity and then added his own: "But that's not really the hard part. What makes this hard is that it is inherently circular. Those mission segments each need start and end weights, which gives you the fuel weight consumed. But to get the takeoff weight for the first segment, you need to know the fuel weight, the payload weight, and the empty weight. But the empty weight depends on how much structure you need, which depends on how much weight it needs to support worst case, which is going to be your takeoff weight.

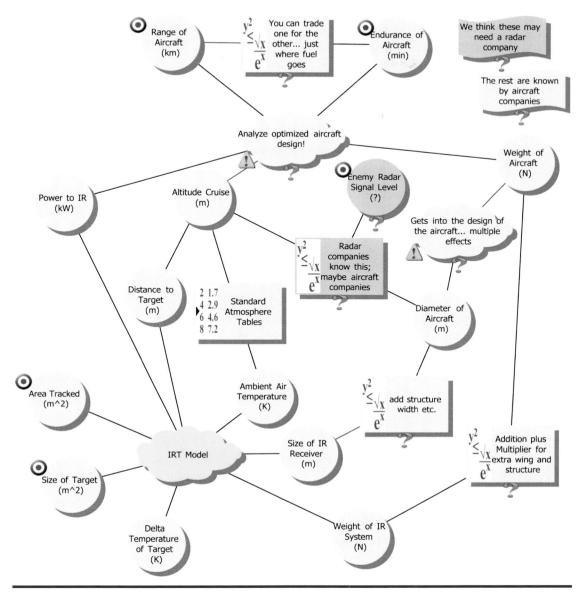

Figure 11.1 Navy Map with knowledge gaps marked.

So, you compute empty weight from takeoff weight. But at the same time, takeoff weight is the sum of the empty weight, the payload weight, and the weight of the fuel required by all the mission segments. So, it is a circular calculation."

Teresa built on that last point: "You need to understand: this is all an inherently iterative process of guessing, analyzing, and revising. All those weights will affect the calculations of each mission segment, which in turn affects the fuel consumed by each segment, which in turn affects all those weights. Does that make sense to you all?" Teresa looked around the room to see if this was registering with anyone. (Most faces in the room had that "I am listening, but please don't call on me to explain what I just heard" look from their school days.) "Sorry if this is unclear; it's complex."

Mark tried to delicately suggest a way out: "We don't want to assume an aircraft design; optimization is important here. But if we started with the UA-243,

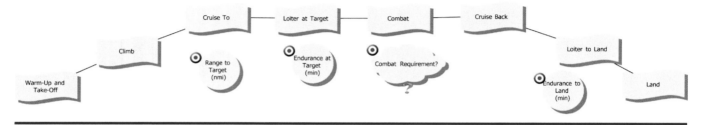

Figure 11.2 Alex's initial Causal Map of the mission segments.

how quickly could you do the analyses? How long would it take to do a few optimizing iterations?"

Laura glared at Mark as Teresa answered, "That depends on what level of fidelity you want in those analyses. Level 0 analyses might take a week or so; Level 1 analyses, a few weeks; more extensive analyses, a few months."

Laura responded, with her glare still directed at Mark, "And so, if the UA-243 isn't very close to optimal for some or all of our target mission profiles, you might end up needing to iterate quite a bit to get to optimal."

Tim, not sure of the source of the tension but sensing a need to defend the viability of the UA-243 that he'd like to sell to the Navy, suggested, "The UA-243 has been designed to cover a wide variety of missions and payloads. Can you give us some clarity on what you're trying to achieve?"

With that cue, Alex stepped in, "Well, let's see if some visual models will help. I've been taking notes in the form of a Causal Map. First, I captured the mission segments. Do I have those right?" (Figure 11.2)

Teresa looked it over. "Yeah, that's what I threw out as a standard mission profile. But in this case we should be checking with the customer. What do you guys say? Is that the mission profile you have in mind?"

Nathan responded, "Yeah, that's the primary mission profile. We could give you general requirements for each of those segments to cover our primary uses."

Alex continued, "Okay, and for each segment you have a start weight and an end weight, which I assume is the start weight for the next segment." (Figure 11.3)

Teresa nodded. "Correct."

"And then you'll have your Breguet Range Equation computing for each?"

"Well, not exactly." Teresa paused to collect her thoughts. "We would for the two cruise segments and the two loiter segments. For the others we'd use different calculations. But what you have pictured there is fine. There *would* be an equation for each."

When papers or books start inserting equations, many of us start skimming past those. So, since we just said *Breguet Range Equation*, many of you may have gone into skimming mode. But notice that we did not actually show the equation itself! Rather, you just see a rectangle representing the equation connected to circles representing the decisions (i.e., variables) in that

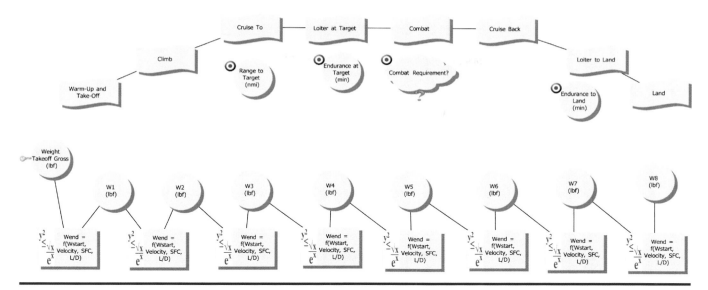

Figure 11.3   **Alex's initial Causal Map with the segment weights.**

equation. The Map is showing you just what you need to know: that an equation between these things is known. The actual equation is just needed by the person or tool doing the computations — and possibly the experts trying to innovate alternatives.

So, please don't start skimming these Maps. Pay attention to the details of how these decisions are related. And how the engineers decompose the larger trade-off relationships into smaller ones until they get to things that are well known.

And that even goes for you manager types as well. Don't assume that it's just the people doing the computational engineering that need to pay attention to the relationships. Everybody involved in decision-making should be able to understand the relationships at this level. Every decision-maker should be challenging the knowledge; only when they have adequate faith in the knowledge can they come to consensus decisions based on that knowledge. And only then will they tend to flush out the knowledge gaps that might otherwise go unnoticed until much later in the process.

And that is one of the key strengths of the Causal Map that we want you to take away from this chapter: that every decision-maker can understand the Causal Map, no matter what area their expertise is in, no matter how rusty their math is, no matter how little time they have available to focus on this area; if they are involved in the decision-making, they should have enough time to at least peruse the Causal Map!

With that much confirmed, Alex continued, "And then I added that the takeoff weight is the sum of payload, fuel, and the empty weight. Where the fuel weight is the total drop across all mission segments. And the empty weight is in turn a

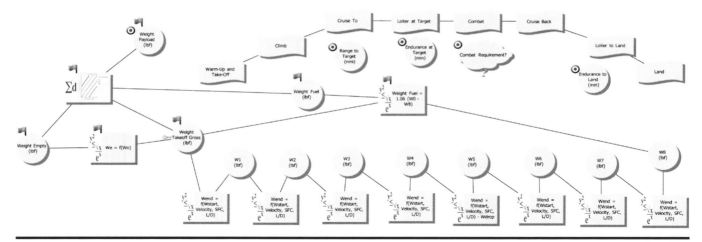

Figure 11.4   Alex's initial Causal Map with the sum of weights.

separate function of the takeoff weight." (See the shapes marked with blue flags in Figure 11.4.)

"Right. And now you can see those two big computational loops that make these analyses so expensive." As she said that, Teresa traced the loops visible in the map shown in Figure 11.4 with her pointed finger.

Again summoning his most assuring voice, Alex replied, "Well, that sounds like a job for a computer. I think our tools will take care of that. But let's not worry about how to best compute that right now; let's just focus on what we know and what we don't know."

With no sense of pushback, Alex continued, "Finally, I added the three other inputs you needed to the Breguet Range Equations: the Velocity, the Specific Fuel Consumption (SFC), and the lift-to-drag ratio (L/D)." (Figure 11.5)

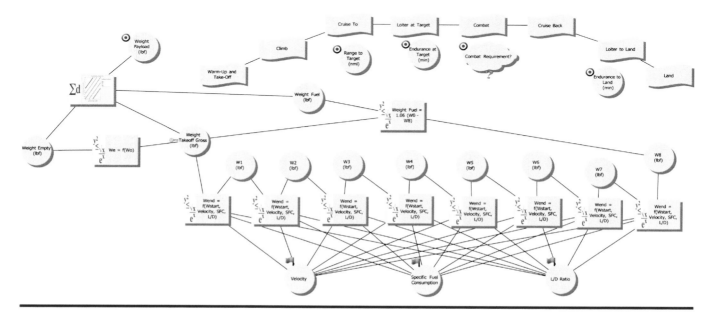

Figure 11.5   Alex's initial Causal Map with the drivers of the range equations.

Teresa responded, "Ah, that's not quite right. Each of those can be different for each segment, which is why we need to break it up into segments."

"Okay, tell me how to fix it."

"What, right now? Do we really want to get into all these details?" Teresa looked around the room to see if she was the only one feeling this was going too deep.

In his typical soft-spoken way, Joe stepped in, answering her question with a question: "Can we answer the questions the Navy is asking without getting into those details?"

"Well… no. But we're not going to do the computations in this meeting; it takes weeks to do a full mission analysis like this. We need to get the designers involved; we'll need to work up some sketches."

Hunting for the best way to keep the conversation going, Joe tried a different angle, "Okay, we may not get to the answer in this meeting, but we want to at least identify what we know and what we don't know so that we can get focused on what we do not know."

Teresa started to dig in a bit, "We know all this; we have tools to help us do it already."

Having seen such objections many times before, Alex made the case: "Except the way you do it requires you to draw the airplane, and that takes weeks. We're looking for a way to bound the problem such that we can avoid needing to draw the aircraft but still be able to use what you know to answer the Navy's questions."

Joe decided to take yet another angle — the "benefit of the doubt" approach. "Bear with us on this. At least this way, the rest of us will gain clarity on why it requires a drawing, but maybe we'll be able to find some innovative way to get at least some partial answers to the Navy's questions."

Having been a silent observer up to this point, Tim decided to weigh in, taking a somewhat different angle: "Alright, I suggest we try it; what else are we going to do in this meeting? Just wave our hands about why it takes weeks? If nothing else, this visual model is going to be a useful communication vehicle — both internally and externally. I am a little intrigued at where this is going."

Scott then offered up his own point of view: "Agreed! If you guys would have given me this picture when I joined UA, I'd have gotten up to speed so much faster. I'm looking forward to getting this filled out more."

Teresa nodded, seeming to agree with that last point. "Okay. Let me think out loud for a minute. To maximize loiter, you'll operate at max L/D; whereas to maximize cruise range, you'll minimize drag, which is generally at a different L/D. Similarly, your SFC for cruise and loiter will be different, and also different from SFC at maximum throttle assumption in combat. For combat, we normally just need to know how long you want to operate at maximum thrust. Oh, and we may drop some of the payload during combat; you might want to split out some of the Payload Weight to be dropped. We won't need the Velocity for loiter. We don't care at what speed we loiter, so we pick the speed that gives us the best loiter time. So it ends up canceling out of the calculation of the fuel consumed for a specified loiter time. We do need velocity for cruise. For takeoff and climb, we do the calculations completely different, and often it's a small enough fraction relative to the whole that we can greatly simplify that. And even more so for Landing; it's pretty easy to descend."

With those adjustments made to the Map (marked by the blue flags in Figure 11.6), Alex decided it would be useful to put it back into the context of yesterday's map (shown again in Figure 11.7). "So, this Map up on the screen is the details that go where the 'Analyze optimized aircraft design' cloud is in the Map we handed out this morning. Logically, it's all really just one Map; but visually, it is often helpful to keep the high-level Map separate, with cloud shapes representing the sub-Maps, rather than merging it all into one even larger Map. But we do need to be careful to check that we have covered everything. We started with 'Range' and 'Endurance'; we clearly have those. We worked our way back to the 'Weight'; we have that covered. But that cloud was also connected to 'Altitude.' We don't have 'Altitude' in our more detailed Map up on the screen. Does that fit in here somewhere?"

"Yes, it does," Teresa responded, "but it is under some of these other numbers like lift, drag, and velocity. For example, the Navy will usually specify their desired cruise velocity, if they have one, as a Mach Number. We need to know the cruise altitude to turn that into a velocity for the Breguet Range Equation."

Alex made those additions (marked with blue flags in Figure 11.8). "Okay, great; one dangling Decision down; by that I mean 'Velocity Cruise' is now

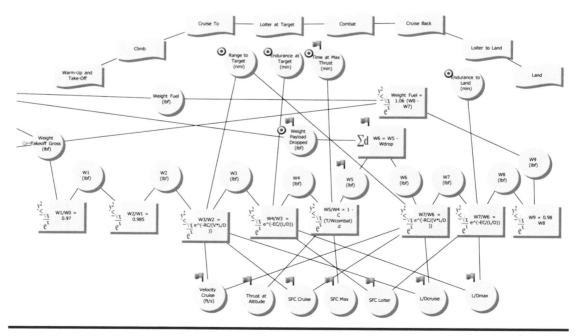

Figure 11.6   Adding the proper inputs to the different segment range equations.

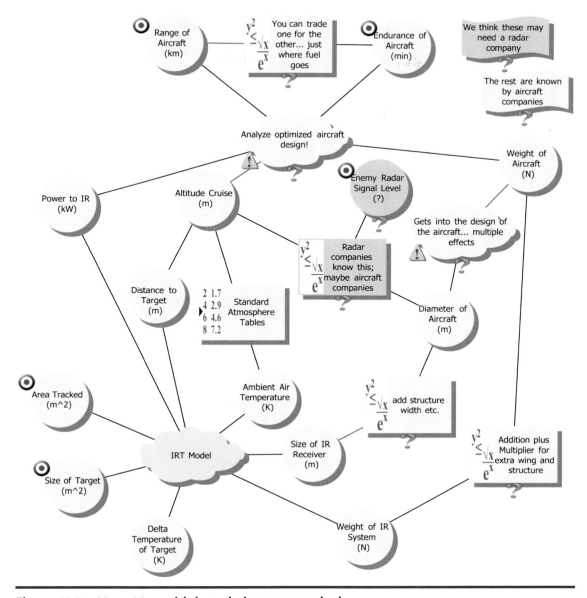

Figure 11.7 **Navy Map with knowledge gaps marked.**

driven by the Navy's requirements. But what about these other decisions: what drives these three SFC values and these two L/D values?"

"All these questions are starting to hurt my head; you're making me think too hard," Teresa sighed. "Well, we get the SFC values from the jet engine specifications or the installed engine performance models. If you know L/D max, you can estimate L/D cruise — or vice versa. But one of the two requires the drawing of the aircraft to start doing those estimates."

"Hmmm…" Debating how to represent the engine in the Map (see the shapes with blue flags in Figure 11.9), Alex asked, "I am guessing the Navy is not going to be dictating the engine to use?"

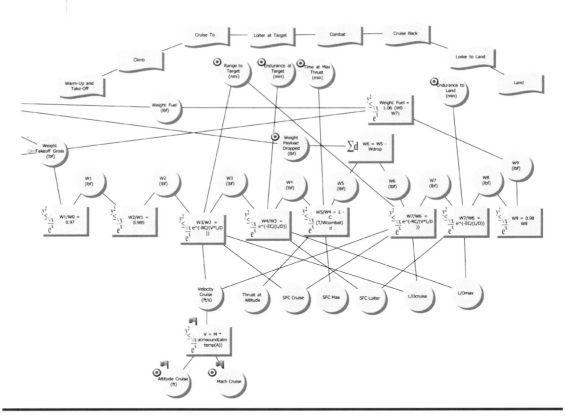

Figure 11.8   Adding 'Mach Cruise' and 'Altitude Cruise.'

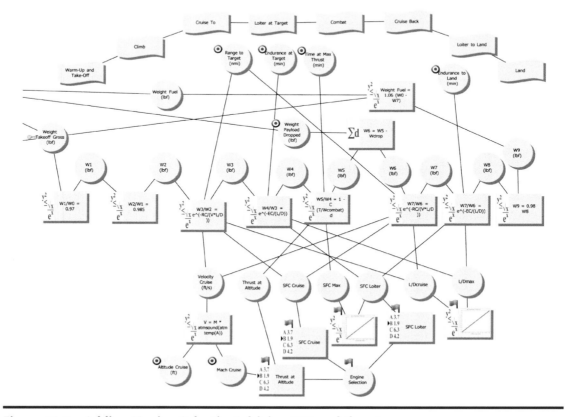

Figure 11.9   Adding 'Engine Selection' driving SFC and thrust.

Nathan confirmed, "No."

"Nor the L/D ratio?"

"No."

"So, those are still what we'd call *dangling decisions*." Alex explained, "There's got to be a customer interest on the other side of these driving them higher or lower, otherwise we can solve this problem real easy: just decide on a really high L/D ratio and an engine with a really low SFC."

After a pause to let that sink in, Alex followed with "Let me guess: cost. You could get a really low SFC, but it would be a really expensive engine?"

"Hmmm." Teresa pondered that a moment. "Theoretically, I guess. But practically, no. Our cost estimates are largely driven by the 'Weight Empty' that you already have in the Map. No, the engine choice, which drives the SFC, depends primarily on the thrust required to takeoff and the thrust required to overcome the drag at the desired maximum velocity or 'Mach Max.' And the drag will get back into the L/D ratios — and the altitude. And the altitude also affects the thrust curves; so, you'll need the 'Altitude' for 'Mach Max.'"

Alex added shapes for those to the Map (Figure 11.10) but paused as he tried to connect them up. "Okay, how do I connect the 'Mach Max' and 'Altitude' to the L/D and thrust and so on?"

"You're getting out of my expertise here," Teresa responded. "We should really get one of our Propulsion guys in here. Or maybe a jet engine company. I'm not sure we can really model the Thrust and SFC as simply an engine selection. This application may call for an engine more optimized for this."

With another sigh at the thought that she was making more work for herself, Laura offered, "We've been working with a jet engine company that uses Causal Maps in their problem-solving. They may be able to hand you a model to drop right in there. Let me give them a call and see if we can conference them into this meeting this afternoon, and perhaps we can get the jet engine knowledge needed here. Should we take a half-hour morning break while I do that?"

"Great!" Thrilled to see the larger IRT strategy playing out surprisingly well, Joe suggested some parallel effort: "Are there similarly some experts that you guys can contact to fill out each box in this Map with whatever equation or data you'd want to use? And fill out each circle with whatever unit of measure and range of reasonable values?"

Teresa, visually walking through the various pieces of the Map, suggested, "Maybe we should reconvene after lunch, say 12:30 p.m.? I would love to go talk to some of my colleagues about some of this Map before we go much further. Alex, can you send me a copy of that so that I can use it as a visual aid with the people I need to talk to?"

"Yes, of course."

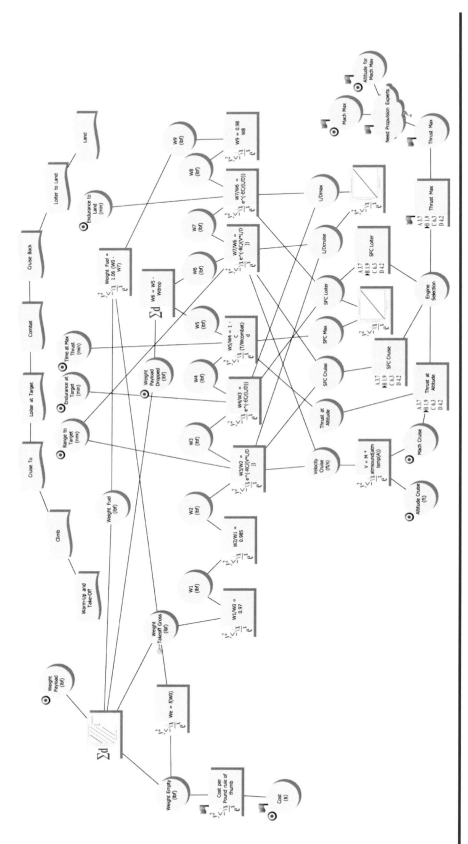

Figure 11.10   Adding 'Cost,' 'Thrust Max,' and 'Mach Max.'

Note how the discussion was enriched and focused by the visual models. We have the opportunity to coach a great many teams in many different organizations on many different types of problems, and we see many of the same patterns over and over. One such pattern occurs when the team reaches the edges of the paper or whiteboard they were given to draw on: they stop drawing and the conversation stops progressing. They keep talking, but their hand waving doesn't create any new clarity, and so the conversation tends to linger not too far off the edges of the visual model created so far. We often let this go on for a bit, just so that we can bring the point home.

"Hey guys, you reached the edge of your paper about 15 minutes ago. Have you made any additional progress in those 15 minutes?"

"Uh… hmmm… not really."

"Go get another piece of flipchart paper and continue mapping what you're talking about."

Let the visual models help pull the discussion where it needs to go.

Also note that as you break things down, you often find more concrete things that you indeed know, and what seemed complex starts to become simpler, while at the same time you continue to find details that you almost surely would have glossed over otherwise. Don't fear the growth of your map; that's just you zooming in, until you are focused on small enough things that they are each simple enough that you can get concrete, and small enough that key details do not fall through the cracks.

Chapter 12

Thursday Afternoon, UA Teleconference with Power Flow Corp.

Laura was able to pull a favor and arrange a 12:45 p.m. call with Power Flow Corporation (PFC), a leading provider of jet engines of all sizes, industry-wide, and one of the largest suppliers to UA. (She chose 12:45 p.m. knowing that meetings never reconvene until at least 10 minutes after the declared time.)

Laura began with introductions: "Rick, Paul, thank you for joining us on short notice. We know you only have 45 minutes, so we can hopefully keep this brief. Rick Moore is one of PFC's senior customer-facing engineers — when there are problems that we need resolved, he is usually the one that helps save the day. In those problem-solving efforts, there is often a Causal Map at the center of the discussion, and often it is Paul Lopez driving, not too different from what Alex has been doing for us today. Rick, Paul, can you see Alex's screen?"

"Yes, we see it."

Nathan then introduced the purpose of the call much as he opened the meeting that morning, just with a substantially larger Aircraft Map in addition to the first day's Navy Map. He walked through the maps, pointing out that there were a couple of clouds the team was looking to replace with the proper details, and the new participants were being asked to provide that expertise.

After some clarifying discussion on the maps, Rick responded, "We do have a detailed model that would fit in there fairly well. But it's more detail than you'd want, and even if you wanted it, we couldn't give it to you; it's some of our most valuable intellectual property — highly proprietary."

The ensuing awkward pause was fortunately interrupted by Paul, who proposed an alternative: "Wouldn't the high-level industry model be sufficient for this? They'd probably prefer to be able to use an existing engine, or at least existing technology, anyway. And that one is approved for release — well, the current state one, not the future state one."

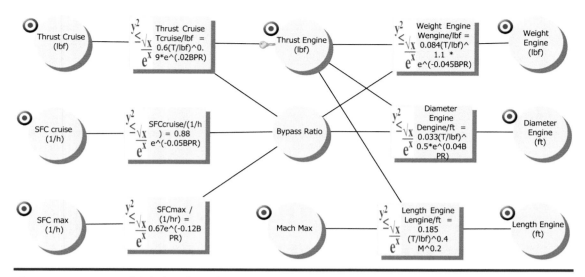

Figure 12.1   PFC's high-level engine map.

Rick pondered that alternative for a bit, looking at how that might fit in where the cloud shape appeared on the Map. "Yeah, I think you're right, that should be sufficient here. Can you pull that up and walk them through it?"

"Sure, give me a sec." Rapid clicking and typing could be heard as he navigated through his data. Paul shared his screen to the web conference, bringing up the Map shown in Figure 12.1, and explained, "Rather than modeling the physics of jet engines, this Causal Map captures where existing optimized jet engine designs have ended up in the trade-off space. The one somewhat internal design decision is the Bypass Ratio. Of course, customers may care somewhat about that if they care about not being detectable by IR devices. So, for some, these may all be customer interests. Not sure I really need to 'walk you through' this Map; it's pretty self-explanatory as it is. Do you have any questions on it?" (Figure 12.1)

Q: What if your key supplier has no experience in Causal Mapping like PFC here? And what if they don't have the knowledge required to help you sort through the details (as UA did when helping the Navy and IRT construct the prior map)?

A: You can build your own model based on your experience as a purchaser of such systems or using industry data of such systems. In fact, this engine model did *not* actually come from a jet engine company. For purposes of this book, we are using only clearly public knowledge. This model was actually developed by Daniel Raymer and made public in his outstanding book *Aircraft Design: A Conceptual Approach.*[1] He developed that model based on data presented in *Jane's All the World Aircraft.*[2]

[1] Daniel P. Raymer, *Aircraft Design: A Conceptual Approach*, Fifth Edition, Reston, VA, AIAA Education Series, 2012, pp. 284–287.
[2] J. Taylor, *Jane's All the World Aircraft*. London, UK: Jane's, 1976.

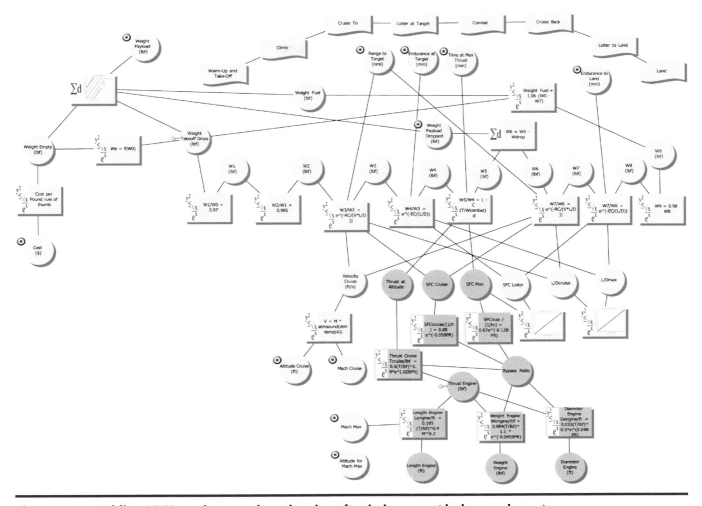

Figure 12.2   Adding PFC's engine map into the aircraft mission map (dark gray shapes).

Perusing this new Map, Alex suggested, "Well, let's drop it into our model and see what questions arise when we see how it fits in."

As Alex made the connections (Figure 12.2), he narrated, "I'll color this engine model darker gray. It connects 'SFC cruise,' 'SFC max,' and 'Thrust at Altitude' to the customer interest 'Mach Max'; that's good. But it does so through relations that add three more dangling decisions that are not customer interests (from our perspective) and don't seem tied to anything else. So, that's not really reduced the number of dangling decisions. But still, that might be exactly what was needed to move this forward, depending on whether we can answer 'What other customer interests will be impacted by changing these dangling design decisions?'"

Teresa thought for a bit. "Well, the engine length and diameter may be of little consequence or may be critical; it depends on how close they are to the aircraft length and the aircraft diameter. If they force the aircraft to be larger, then that will affect weight, wetted area, drag, and the L/D ratio."

"Okay, have I captured this right? Is this L/D the same as one of the L/Ds that we already have in the Map?" (Figure 12.3)

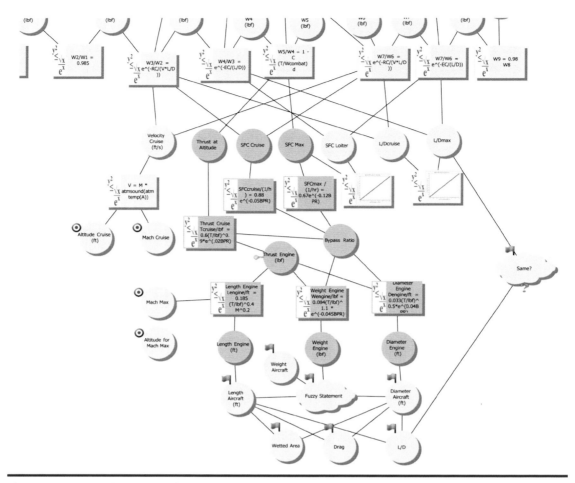

Figure 12.3   Adding aircraft geometry to accommodate engine geometry.

"Yes, I am talking about the 'L/Dmax' that you already have," Teresa responded, "but it's really that 'Length' and 'Diameter' go into 'Wetted Area,' 'Wetted Area' then affects the 'Drag,' and that goes into the 'L/Dmax.' But I'd need to think through how to model that here, because the 'Drag' will depend on velocity and altitude. I think we want to tie 'Wetted Area' directly to 'L/Dmax,' but I'll need to work with my guys to figure that out."

"Okay, I'll use a cloud for that, indicating that several things may go there, and mark that as a knowledge gap to be closed. So, is this right now?" (Figure 12.4)

Teresa took a moment to look at the various shapes and question their connections. "Well, 'Wetted Area' is also going to include the wing area. I think we'd want to use what we call 'Sref,' the planform reference area. We can work up a reasonable formulation for that as well."

"Okay, I'll add a cloud for that as well. So, we seem to have the engine model in dark gray completely connected into the larger map." (Figure 12.5.) "But before we let Rick and Paul go, I'd like to ask them, is there anything else that you see in this Map that would have an impact on the jet engine performance, weight, cost, and so on?"

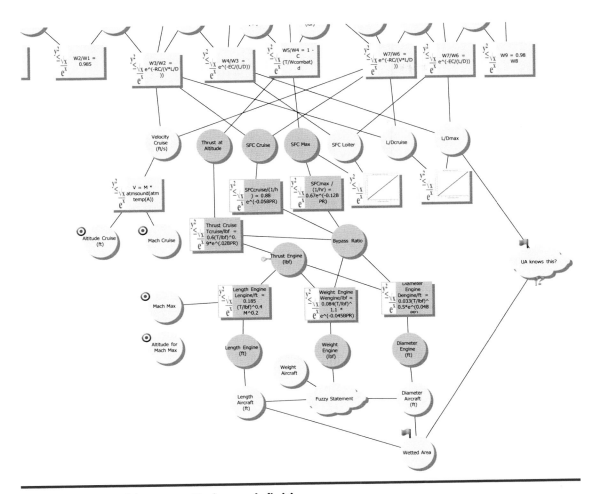

Figure 12.4   Revision to wetted area definition.

After a bit of a pause, Alex added, "And while they think on that, does anyone here see anything in the dark gray that might impact other things in the Map or other things that we care about that should be in the Map?"

After a bit of discussion, the group concluded that all of the important impacts were captured in the Map, and the pair from PFC were thanked and exited with plenty of time to spare.

With that, Joe wanted to return to the first day's Map (shown again in Figure 12.6) and again reevaluate whether the revised detailed Map in Figure 12.5 covered everything connected to the cloud it represented.

Joe asked, "How about this 'Weight of Aircraft' from the Navy Map: is that one of the many different weights we have up here in this Map?"

"Hmmm… Yeah, logically, all of that is part of 'Weight Empty,' but…" Teresa answered with a furrowed brow, "but we are already accounting for all that in our computation from the takeoff gross weight. We could perhaps factor that in with some far more sophisticated model. My suggestion is that we do *not* do that here."

"So, should we drop these shapes I've marked with red Xs here?" as Alex marked each shape he was suggesting to delete (Figure 12.7).

Teresa concurred, "Yeah, I think so."

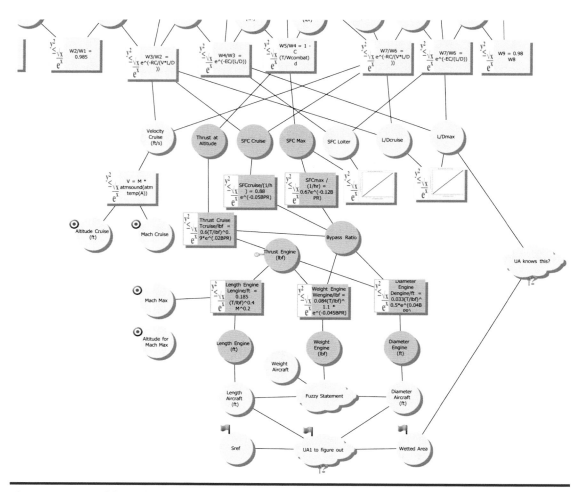

Figure 12.5   Adding the reference area ('Sref').

"Okay, deleted." Alex deleted the marked shapes. "That just leaves 'Sref' dangling here."

Teresa pondered what was impacted by 'Sref.' "Well, 'Sref' goes into almost everything. In fact, rather than using 'Weight,' we often compute everything in terms of 'Wing Loading,' which is 'Weight' divided by 'Sref.' But since we already have 'Weight' here, and we're assuming the math will somehow work out, I don't think we need 'Wing Loading' here."

Alex then deleted the 'Wing Loading' shapes he had just added as Teresa was talking. "So, what here does 'Sref' need to go into? Or perhaps the better question is this: What is missing from this map that would limit 'Sref'?"

"Let me think." Teresa leaned back in her chair as she pondered that question. "The 'Aspect Ratio' is 'Wing Span' squared divided by 'Sref', and 'Aspect Ratio' is limited by what will work for different types of aircraft." (Figure 12.8.)

"Types of aircraft?" Alex asked.

"For example, Fighters versus Bombers versus Cargo planes versus Transports. For each type, there's a certain maximum 'Aspect Ratio' that would be reasonable; above that, the aircraft generally wouldn't be able to

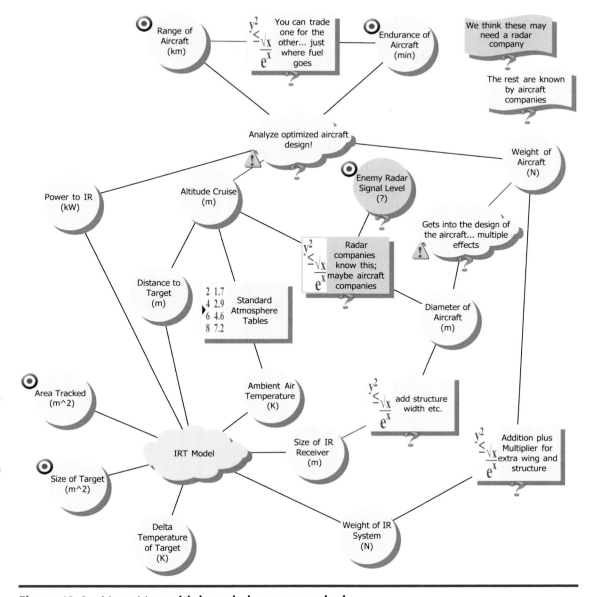

Figure 12.6   Navy Map with knowledge gaps marked.

perform as needed. We use that as a key guideline when we start drawing airplanes."

Verifying his understanding, Alex responded, "Ah, so that's really rule-of-thumb data that's capturing what's needed to satisfy a collection of different customer interests associated with each aircraft type. For example, a Fighter would need much higher maneuverability, I would assume."

"Yes, exactly."

Alex elaborated, "That's a nice simplification that you currently use and which we can similarly use in this Map." (Figure 12.9.) "However, we should remember that it is a simplification. If later we need to make visible trade-offs with some of those specific customer interests, then we may need to break those out. In other words, not handle them in terms of simply aircraft type."

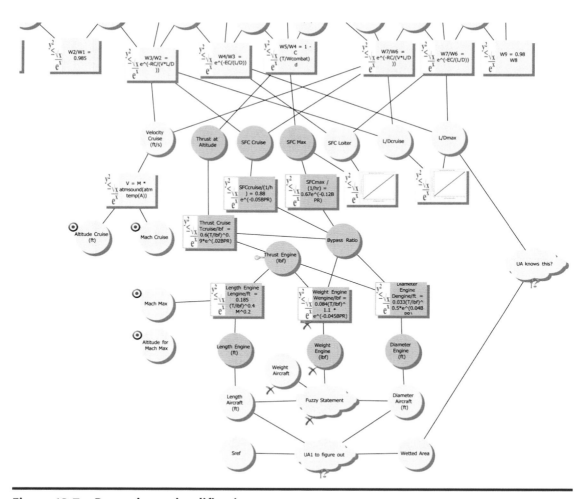

Figure 12.7   Proposing a simplification.

Teresa nodded. "Good point."

"So, how do you model that rule of thumb?" Alex asked.

"Let's see." Teresa massaged her forehead. "We have a couple tables of values associated with each type: one a multiplying factor, the other an exponential factor. And then there's an equation that puts them together to form an upper bound on 'Aspect Ratio.'"

"Great, I've captured that into the map." (Figure 12.10.)

Alex continued to probe: "So, now we have two dangling decisions: 'Wing Span' and 'Aircraft Type.' What else do those impact?"

After a long pause, Tim jumped in: "Hey, while Teresa ponders that, perhaps it will help to add some of the knowledge we gathered during the break-out before lunch. I texted one of our aero guys a screenshot of that cloud on the L/Dmax. He sent me a couple different rough calculations for that. One of them is based *not* on 'Wetted Area' but rather 'Wetted Aspect Ratio'; so that would tie 'Wing Span' back in there. It has the same formula as 'Aspect Ratio' but with 'Wetted Area' instead of 'Sref.'"

"Great, got that." (See right-hand side of Figure 12.11.) "Any other knowledge gathered this morning?"

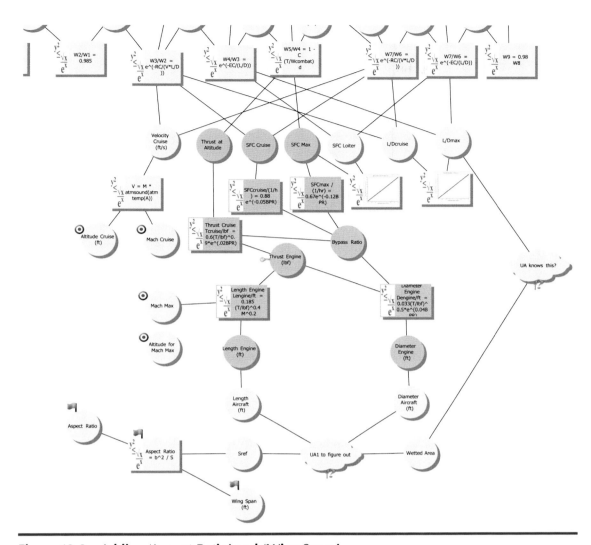

Figure 12.8 Adding 'Aspect Ratio' and 'Wing Span.'

Scott added, "I walked one of my testing colleagues through the Map over lunch, asking which boxes we had data or equations for. He is pulling that data now. But he asked if it would make sense to add a limit on 'Fineness Ratio' here? We know the optimal 'Fineness Ratio' ('Length' divided by 'Diameter') will be between 6 and 8 for a subsonic aircraft. He thought that should be useful in limiting the design space."

"Good point." Teresa turned to Alex. "But can we add additional limits to things that already are tied together?"

"Absolutely. You can add as many different limits that exist; the more knowledge about these things, the better." (See the added shapes with blue flags in Figure 12.12.)

Scott nodded approval, but skeptically questioned further, "If you have two equations computing the same thing, which one wins?"

Alex smiled as he saw their eyes starting to open to the impact of Set-Based analyses. "Remember, we are not computing a particular value — this isn't a

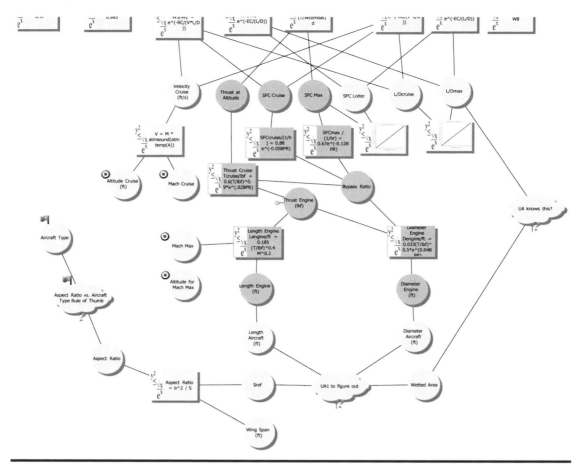

Figure 12.9   Adding 'Aircraft Type.'

point design — we are computing the sets of what's possible. So, it will be the intersection of the sets computed by each of the relations."

"Ah, that makes sense," Scott answered. "Not sure how you do that, but it's logical."

Note the valuable additions from having other experts' eyes on the Map. It is in complex decision-making scenarios where you need people with different expertise collaborating that the Causal Map truly starts to shine — but only if you get up from your chairs, leave the room you are sitting in, and get that Map in front of other experts (or teleconference those experts in)! In traditional meetings, we rarely think about moving the meeting forward by taking an extended break. That's where three of the letters of LAMDA come in; we want you thinking, "Can we go *Look*?", "What else should we be *Asking*?", and "Who else should we be *Discussing* this with?"

At one of our clients, during the first morning of a two-day Set-Based Thinking® workshop, the answer to "Can we test that somehow?" (a variant of "Can we go Look?") was "Well, we do have a prototype in the shop. We could just _____ and that would give us an upper bound on that." Given it could be

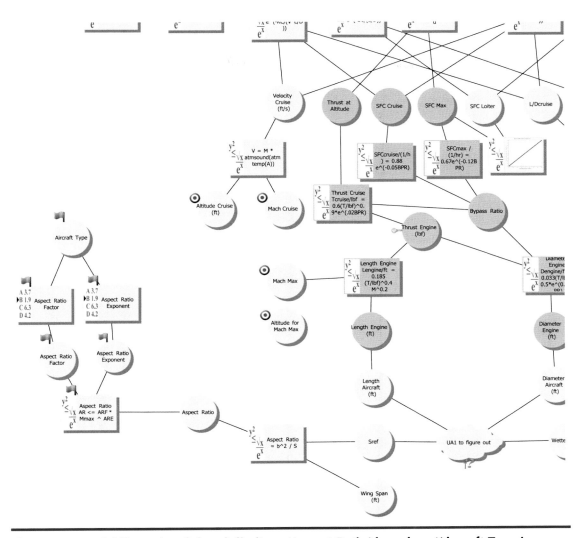

Figure 12.10   Adding rule-of-thumb limit on 'Aspect Ratio' based on 'Aircraft Type.'

closing a critical knowledge gap, we decided that would be the most valuable use of the afternoon. That knowledge greatly focused the analyses that we did during the second day of the workshop.

Don't forget to think LAMDA. And don't be afraid to take a break from the meeting to act on LAMDA.

Alex continued the probe: "Great, so are there other things that might be impacted by 'Aircraft Type'?"

"I am sure there are," Teresa responded, "but I am not seeing anything here. Let me think."

After a bit of a pause, Tim asked Teresa, "You said you have a number of tables of rule-of-thumb or historical data driven by the 'Aircraft Type.' Can you pull those up? Perhaps those will tell us what's missing."

"Oh, that's a good idea. During the break I discussed this with one of the guys who does the initial sketches of the aircraft. I asked him about his rules

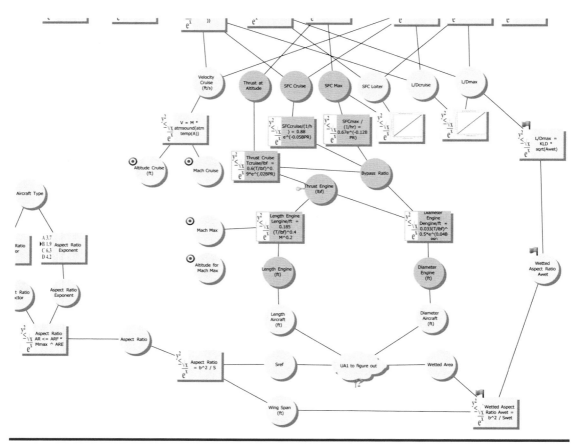

Figure 12.11   Adding 'Wetted Aspect Ratio' driving 'L/Dmax.'

of thumb. He said he'd email me what he uses." Teresa quickly pulled up a few different files, all keyed one way or another by the "Aircraft Type." Some were PowerPoint files with charts for each type, some were scans of what looked like documents from the days of typewriters, some were Excel worksheets. "Well, one of these is just like how you have mapped the 'Aspect Ratio.' We compute a lower bound on the length of the aircraft based on a factor and exponent, each based on 'Aircraft Type.'"

"Okay, let me make a little room for that and duplicate these." Alex copied the 'Aspect Ratio' shapes, changing their names to 'Length' (Figure 12.13).

Teresa continued looking through the documents. "Oh, this one will be useful! There's a table of 'KLD' values for each 'Aircraft Type.' 'KLD' is the multiplying factor in your relation that is computing 'L/Dmax' from the 'Wetted Aspect Ratio.'"

"Oh, I thought that was a constant. I should have had a decision shape for 'KLD.' Let me add that and connect it up here."

"Ah, I see," Teresa said to herself as she waded through the many worksheets in the Excel workbook. "There are a lot of these sheets because they are multi-dimensional lookup tables. But really this is all just one equation from these two lookup tables: again, a factor and an exponent. But this time the equation doesn't just limit one decision, it limits the relationship between

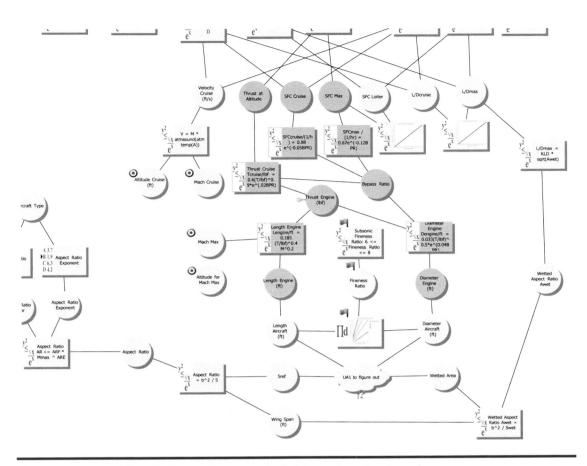

Figure 12.12   Adding limit on 'Fineness Ratio' for subsonic aircraft.

'Takeoff Gross Weight,' 'Mach Max,' and 'Engine Thrust.' This is capturing their rule of thumb for the minimum thrust they will need."

"Okay, let me rearrange things a little bit here. Does that look right?" (Figure 12.14)

Scott commented, "Wow, this looks really good. I think I actually understand this better now than I ever have."

Teresa concurred, "Yeah, I already knew all this; but it is now far better organized in my mind."

Note that not all relations in your Map need to be the fundamental physics. Keep in mind the goal here is not to model the physics but rather to guide your decision-making, — to model your best practices for how your organization makes the decisions that it needs to make. As such, relations such as those surrounding the "Aircraft Type" decision that are capturing where your optimized and corrected designs have historically ended up can be incredibly valuable.

Further, note the power of Set-Based in that. Such historical numbers can't give you point designs, but they can give you upper and/or lower bounds where things tend to work well versus where you start taking on a lot more risk.

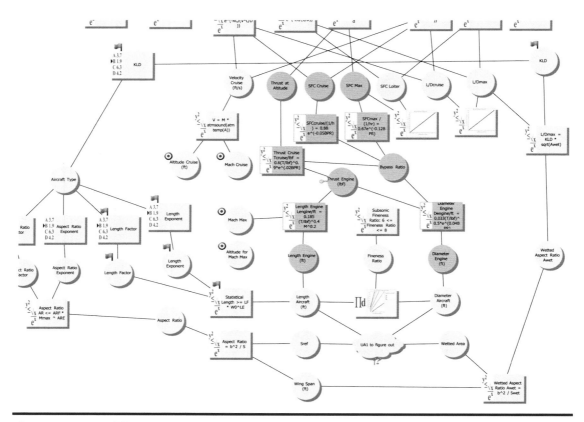

Figure 12.13   Adding rule-of-thumb limit on 'Length of Aircraft' based on 'Aircraft Type.'

A lot of the knowledge that is "in people's heads" that we seem to have great difficulty capturing and making reusable is of this form. Our "rules of thumb" and our "gut feel" tend to be fuzzy ranges where we feel comfortable based on years of experience working in a design space. It is this network of Set-Based upper and lower bounds on what tends to work versus what tends to cause problems that our Causal Maps are able to capture in a way that can directly connect what you know to what you need to decide. That is what makes our Decision Maps so reusable.

And again, where these Maps really start to shine is when the problems are complex enough that you need to somehow bring together multiple different experts' rules of thumb, historical optimizations, and so on. To allow collaboration in terms of such knowledge, you have to make that knowledge visual in such a way that it can be interconnected; that is the role that these Maps play.

Joe requested, "Let's look back again at our cloud in the Navy Map and see if we have everything covered."

Alex started using a green check on each shape in the Navy Map (Figure 12.15) that was covered by the UA Map (Figure 12.14). "Actually, we have both of these aircraft clouds covered now because we have the relationship between 'Diameter of Aircraft' and 'Weight,' along with 'Weight of IR System' as 'Payload Weight.'"

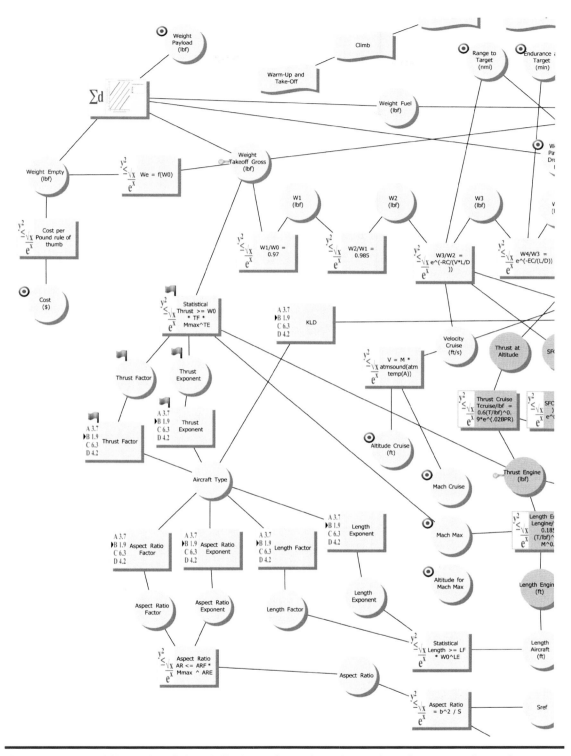

Figure 12.14   Adding rule-of-thumb limit on thrust based on 'Aircraft Type' and 'Mach Max.'

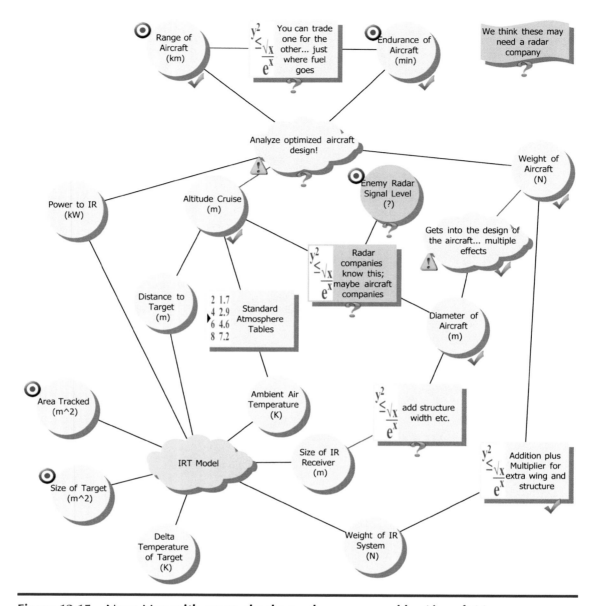

Figure 12.15   Navy Map with green checks on shapes covered by Aircraft Map.

Except for one very key thing: 'Power to IR.' What is the impact of using the engines to power the Hawkeye system?"

Teresa cringed a bit. "Well, if that will require additional thrust from the engine, then that might invalidate all of these rules of thumb! We might need a propulsion expert to work with us on that."

Tim offered up a few simplifying assumptions: "A question for you Navy guys: will the Hawkeye system ever be needed during takeoff or during the combat phase, which in your case I guess means 'an avoidance maneuver and run' if attacked?"

Nathan answered, "Definitely not during takeoff. And if it was slowing down an escape, we'd definitely want to have a quick cutoff. Nor during Cruise Out or Cruise Back. Just during Loiter at Target."

Tim continued, "Okay, then I think that greatly simplifies things. During 'Loiter at Target,' we are using very little of the 'Engine Thrust.' As long as the Hawkeye requires less thrust than takeoff or maximum Mach, then it won't affect the calculations here at all. Am I thinking about this right, Teresa?"

"Yes, exactly. In fact, our engine performance models are very effective in predicting the impact of power loads on engine efficiency. So, we can pretty easily estimate how much fuel will be consumed based on the required 'Power to IR' and 'Endurance at Target.' And since that is independent of the Breguet Range Equations, I would just model that extra fuel weight as the 'Payload Dropped,' since there are no weapons being dropped here."

"Great, so we have everything covered now," Alex asserted, looking around the room for confirmation (Figure 12.16).

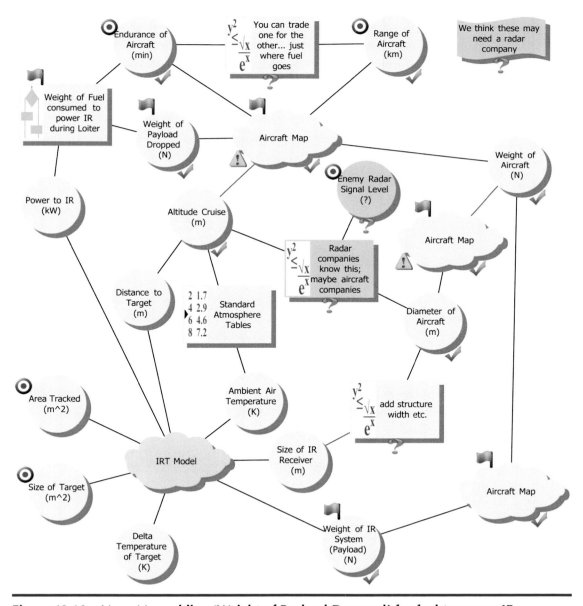

Figure 12.16   Navy Map adding 'Weight of Payload Dropped' for fuel to power IR.

Although she knew she was going to get a negative answer, Laura still asked, "So, does that mean that you can now answer the questions that we asked?"

Alex smiled, "Well, the Map is fully connected. But we still have a lot of details to fill into each of these boxes. Once we do, though, what we will be able to do is not just give you an answer to your questions but show you the set of all possible answers to your questions. And then you can choose whichever of those is most appealing to you. You get to optimize that as the customer."

"Okay," Laura said skeptically. "Not sure I understand how that's going to work yet, but that sounds very appealing. What are the next steps? And how quickly can we do those?"

Alex looked back at the Map. "We need to verify the units of measure and ranges of values on each of these circles — each of the decisions. We need to gather the equations, lookup tables, rule-of-thumb data, or whatever it is we are going to use to compute each of these boxes — each of the relations. And as you visit the people who know each of these things, I suggest you ask them to evaluate the Map. Try to identify anything that is missing. Ask them, 'What else might be impacted by these decisions? What else might constrain these decisions?' Almost surely, we will end up needing to modify this Map as we fill in the details."

Teresa nodded. "Yeah, I definitely want to run this by several of our designers who do the initial aircraft sketches. They each use their own rules of thumb and each follow a somewhat different logic, plus use a fair amount of their own human judgment, in coming up with their initial drawings. It will be interesting to see what suggestions each makes — or what objections each has. As far as timing goes, Laura, I think I can probably get this done tomorrow and Monday. We could then get together Tuesday, or to be safe Wednesday, to incorporate that into the Map." She turned to Alex. "Then what?"

"Then I'll need a day or two to do the calculations and generate Trade-Off Charts showing the feasible design space — the sets of possibilities. So, perhaps the following Monday? You might want to plan for Monday and Tuesday. There's a good chance you'll want to see things differently than I first produce them, as this is a complex multi-dimensional design space. The Wednesday meeting next week could be done virtually, but the Monday–Tuesday meeting the following week would be best in person. We'll usually end up drawing charts and timelines and such."

Laura asked, "Does that work for everybody?" Though given her tone, most in the room heard, *"That works for everybody, right?"*

"Oh, one bit of homework for you." Alex looked at the Navy group. "We have a knowledge gap on how you want to specify 'Radar Signal Level' and how that is related to 'Altitude' and 'Diameter of Aircraft.' Can you close that gap by Wednesday?"

Mark answered, "Yes, I should be able to get an answer on that by then."

Chapter 13

Wednesday Morning, Week 2, Teleconference between the Navy, IRT, and UA

As all the attendees of the prior meeting started showing up in the web conference's Participants list, Alex shared his screen, where he had queued up the K-Brief containing the two Causal Maps that they had been working on. Teresa was reviewing her notes and deciding in what order to present them to the group for incorporation into the growing Maps.

Stealing the role from Nathan, Laura opened the meeting: "I scheduled this for three hours; I'm guessing we'll adjourn well before then. But I believe the objective is to close all the knowledge gaps that we were each working on since our meeting last Thursday, such that Alex has all the knowledge he needs to answer our questions on Monday. Do I have that right?"

Alex replied, "That is the plan. Can everybody see my screen? Who wants to start?"

Teresa took the lead, "Alex, I emailed you five pages with equations written out so that I would not have to speak them to you over the call."

"Ah, yes, I got those. I'll pull them up."

Teresa continued, "I spoke with three different senior aircraft designers; they each gave me some additions to the Map. The first is another factor plus exponent pair coming off 'Aircraft Type' like the others — but this one limiting the 'Weight Empty' calculation from the 'Takeoff Gross Weight.' Two of them also make a small adjustment to that calculation based on whether the design has variable sweep in the wing design. If so, they bump up the weight by 4%. All of that is documented on the first page."

"Let me rearrange the Map a bit to make some room for that here." Alex dragged some of the shapes around (Figure 13.1). "Okay, I think I have that captured."

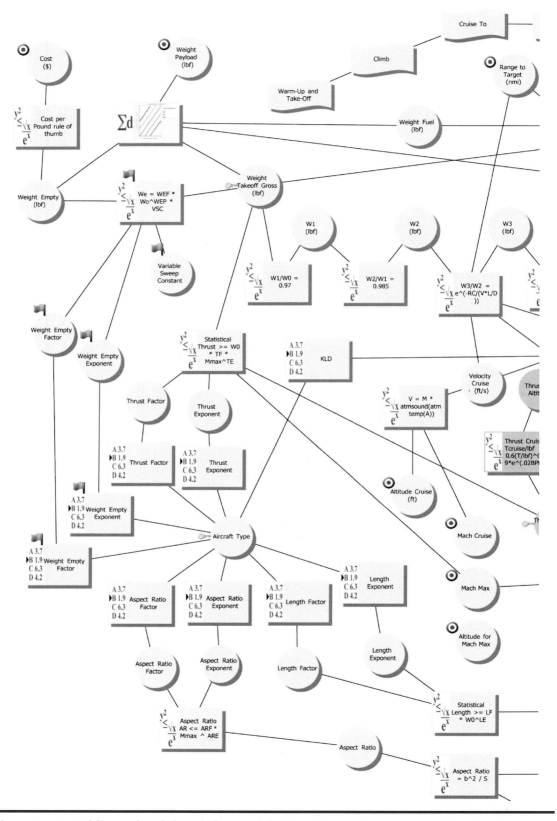

Figure 13.1   Adding rule-of-thumb for 'Weight Empty' based on 'Aircraft Type' and 'Takeoff Weight.'

Teresa continued, "On the top of the second page, I have the equation for computing 'Wetted Area,' which we call 'S-wet,' to replace this cloud at the bottom with the question mark."

"Great, that's easy to add to the Map."

"The rest of this page will not be so easy. A couple of the designers were uncomfortable with what we have there for 'Diameter' and 'Length.' The real limits, they think, will be from the net volumes of the aircraft. Everything is fighting for space with the volume of fuel needed to achieve the 'Range' requirement. That is a very direct trade-off. So, from the total weight of the fuel, we can compute the required volume. The net volume available for that fuel can be estimated from the length and diameter of the aircraft subtracting out the volume of the engines computed from their length and diameter and the volume of the payload. They have given me a few different ways to compute that, which I have combined into what you see on the page."

Alex looked it over. "Okay, the Diameters of the engines and payload have to fit. The lengths of the engines and payload have to fit. And the volumes of both need to leave room for the fuel volume. And I see they have rough padding factors for each. And then 'Volume of Fuel' is computed from 'Weight of Fuel' based on 'Density.' That all makes sense."

Teresa continued, "One of the designers gave formulas for stall speed; however, two others argued that would not be an issue in this case. Should we add that here?"

"I will add a cloud as a placeholder so that we remember to revisit that question after we have narrowed the design space."

"Two of the designers mentioned maneuverability and suggested some turn rate calculations off 'Mach Max' and 'Max Lift' and such. However, a question for Nathan and Mark: I don't think you want any sort of dogfighting capability here, right? You just plan to turn and run if detected?"

Nathan replied, "You are correct. But let's do like we did with the Stall Speed and put a placeholder cloud in for that. The 'turn' in 'turn and run' could be important, depending on how high of an altitude this design ends up at. But yes, my guess would be that it is a non-issue." (Figure 13.2.)

Teresa continued to explain her learning: "The third page has the final thing we need to add, and it is something that all of the designers gave me grief over: the real driver for the engine thrust did not seem to be reflected properly. The required 'Engine Thrust' will more likely be driven by the takeoff requirements than just 'Mach Max,' but all we seem to be reflecting is 'Mach Max.' We have the 'Takeoff Gross Weight' in here, but other than that, nothing on takeoff requirements. I know we discussed takeoff requirements being the driver, but I couldn't find where we captured that on the Map."

The whole group agreed with Teresa's sentiment there: they all remembered discussing it, but none of them noticed that it didn't make it onto the Map.

Alex shared his past experience: "Yeah, that's one of the big benefits of the Causal Mapping process. It is so easy when discussing complex problems to

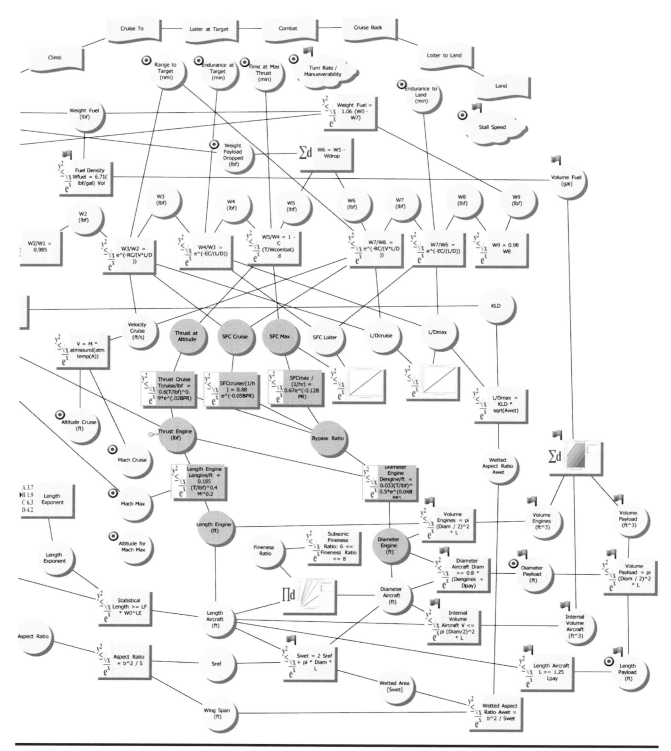

Figure 13.2 Adding geometric relationships and limits on volume of fuel.

focus on one part of the information that's being discussed such that you fail to notice that another part of the discussion is not being given the proper attention. Because it was mentioned, you get a false sense of coverage. By having multiple eyes on the growing visual Map, you make sure nothing is forgotten or glossed over. Those fresh eyes don't see what you talked about, they only see what made it onto the Map or other visual models. So, they tend to catch things."

So, how many of you Sherlock-like sleuths picked up on the prior mentions of takeoff being the key driver of 'Engine Thrust' but that it was not being reflected at all in the growing Causal Map? If you did catch that, we salute you (and recommend you play Alex's role in your teams)! Of course, many of you may not really be studying these Causal Maps deeply and thus have an excuse for not noticing it. That is understandable given you do not need to design an aircraft, but we hope you recognize the importance of truly examining these Maps deeply when your teams start to develop their own Maps.

Consider for a moment if mystery books/movies came with a Causal Map that incorporated all the facts and displayed all the causal relationships. It would eliminate a lot of the mystery and surprise twists, would it not? Isn't that what you want for your product development projects (no surprise twists)?

How expensive would it have been if the key driver, the takeoff constraints, had been forgotten until after all the trade-off decisions had been made and lots of CAD work had been done. You might think, "They would have thought of it before then." However, there are countless stories of engineering disasters where exactly such things have happened (e.g., not factoring in the weight of the books into the structural design of a second-floor library).

"So, what is it that I need to add to the Map to drive 'Engine Thrust' from 'Takeoff Gross Weight'?"

Teresa replied, "There's actually several customer interests regarding takeoff that we are missing."

"True," Nathan interjected. "Ideally, 'Takeoff Distance' would be less than 300 ft to avoid the need for assisted takeoff on the aircraft carrier. But that is not a strict requirement, as assisted takeoff is possible."

Laura agreed, "That decision would be part of the optimization process; we definitely would want that, but might not be willing to accept the trade-offs to get it."

Teresa continued, "In general, in addition to 'Takeoff Distance,' we need to know the 'Takeoff Clearance' (how high we need to be at the end of the 'Takeoff Distance') and the 'Takeoff Altitude,' since that affects the thrust, lift, and drag computations. On this third page are the equations for the three Relations you'll need to add."

"Great, let me connect those into the Map. I'll capture the actual equations later." (Figure 13.3.)

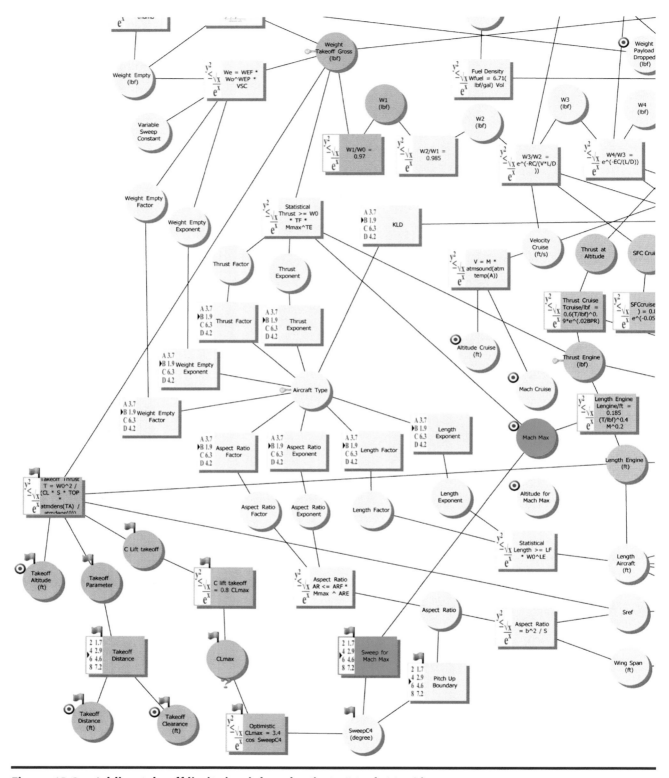

Figure 13.3   Adding takeoff limits in pink and a tie to 'Mach Max' in green.

"I have colored all the things related to the takeoff stage pink and the stuff related to the combat stage green. As Maps grow, I often find some visual groupings with color (whether with colors of pens or sticky notes) can help."

Teresa nodded in agreement, "Yeah, I think that helps."

"What else?" Alex asked.

Teresa responded, "On the last two pages I've put the data tables and equations for each of the other relations. Other than that, my team thinks this Map is pretty complete for this level of analysis."

Mark volunteered his homework: "I did get a formulation for 'Radar Signal Level': 'Sref' divided by 'Altitude' to the fourth power, which makes the units 1/ft² (or 1/m²)."

Alex pulled the Navy Map back up and updated the 'Radar Signal Level' shape (Figure 13.4). "I need to add 'Sref' in here instead of tying it to 'Diameter

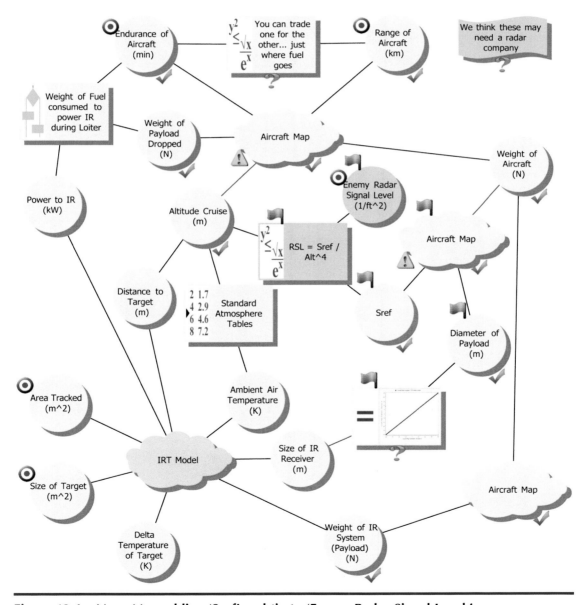

Figure 13.4   Navy Map adding 'Sref' and tie to 'Enemy Radar Signal Level.'

of Aircraft.' And actually, 'Size of IR Receiver' should just tie to 'Diameter of Payload,' as the Aircraft Map then ties that to 'Diameter of Aircraft,' and to the Volume calculations."

After a bit of a pause, Alex observed, "Hmmm, those Volume calculations that we added to the Aircraft Map also impact this Map. From 'Weight of Payload Dropped,' which is Fuel, we'll want to compute 'Volume of Payload Fuel' and use that to compute 'Volume of Payload' in the Aircraft Map. We won't have a hard 'Payload Length.' We can leave out that Relation for the Navy analysis. Hmmm, I need to rearrange this Map a bit to get the Diameter over with the Fuel calculations to compute 'Volume of Payload.'"

Mark corroborated Alex's changes to the map (Figure 13.5): "Yeah, that looks right. However, I should point out that when I asked for what values

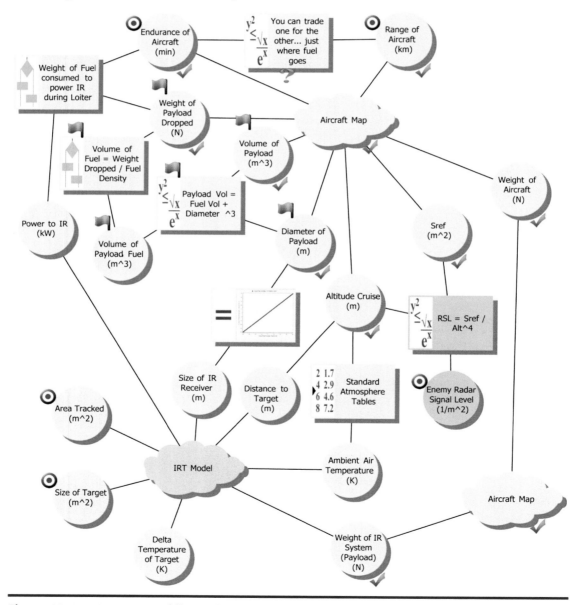

Figure 13.5   Navy Map adding volumes and ties to weights.

of 'Radar Signal Level' they would desire, since it is something we would be targeting, I got some pushback. Some argued that most 'customers' of this wouldn't be too familiar with the 'Radar Signal Level' values and would actually have a better feel just driving the 'Altitude' numbers. Others pushed back that 'Radar Signal Level' is just part of whether it is detected and that detection is not even the key issue. The higher the altitude, the more difficult it is to do anything about it, even if they detect it. The real issue is *survivability*, and there's a number of things that go into that, some of which are actually government classified. However, they asserted that as long as we can keep this at a high enough 'Altitude' — say, above 60,000 ft — then most of those details could be ignored. But in any case, the survivability will need to be evaluated by a separate group and confirmed, once we have adequately narrowed the design."

Joe responded, "Such complexities are not uncommon. One of the big advantages of working Set-Based is that you can explicitly do just as those experts were suggesting: design your way around those complexities by essentially eliminating that part of the design space as 'weak' because of those complexities. However, you remember you did that, in case you run into other limits that make it desirable to investigate those uglier parts of the design space. So, for now at least, I propose we take the suggestion to work in terms of 'Altitude' and see if the trade-offs will allow us to easily keep 'Altitude' over 60,000 ft to minimize the risk that we encounter any of those survivability issues. That sounds like an area we really don't want to step into if we can avoid it."

The group concurred with that recommendation. Teresa added, "So often when we try to get into detailed analyses such as these, we get mired in details where we lack knowledge. I really like how this process leverages the knowledge we can easily get to help us work around the knowledge that we cannot easily get."

Joe echoed, "Yes, leverage what you know to design a system robust across the full range of values that you don't know."

Alex then asked, "Okay, any other knowledge that we need to incorporate here?"

Mark, always a stickler for details, responded, "That all looks good, but there's one question mark left up top. Does that need to be filled in?"

Alex replied, "Good point. That represents the knowledge gap in how you, the Navy, will want to make that trade-off between 'Range' and 'Endurance.' That's actually a trade-off decision that you can make as late as when you launch the mission. So, it could be argued we should delete that here. However, if we're going to leave both 'Endurance' and 'Range' as customer interests in the Map, you need to realize that you can trade those off in either direction you want. In other words, this design, with those Decisions in it, will always be a set of possibilities, not a single point. That will be important to keep in mind as we start looking at the Trade-Off Charts."

"Ah, that makes sense," Mark replied. "It will be interesting to see how that plays out in the Trade-Off Charts. I'm not sure I can visualize that yet."

With her now-standard skeptical look, Laura added, "Yeah, I am definitely looking forward to seeing how this allows us to optimize the design for our needs. Like Mark, I am not sure I can visualize that yet. But I do find it appealing that the things we can trade off right up to mission time are being made visible and that we might be able to see the design space that we will have available to us at mission time!"

"Alright," Alex concluded, "I think I have everything that I need! I'll connect this around our model (the 'IRT Model' cloud) and then use that to make visible the design space and trade-offs, and I'll have things to show you Monday morning."

Nathan noted, "Definitely didn't need all three hours, Laura. Signing off." There followed a symphony of goodbyes and see-you-Mondays from the group, as the entries in the participants list of the web conference quickly vanished.

> The observation that the design may not converge to a point until mission time (that some of the decisions can be delayed until the launch of the mission) extends to all phases of the product life cycle. For example, some decisions may be left open for manufacturing to decide. Rather than applying tight tolerances to things in design and then working hard to get stamps or similar operations to hit those, Toyota would design for sets of possibilities such that as long as the initial tooling produced parts that were within the sets and matched properly, they could be approved and those decisions effectively made during the tooling production. The cost and time efficiency of that approach versus demanding they hit very precise tolerances was huge but still allowed them to have television commercials touting how precisely matched the seams of their Lexus cars were.
>
> Once your design teams become accustomed to designing in terms of sets of possibilities, they can begin to leverage that ability as an additional tool for optimizing the larger supply chain and value stream, by loosening specifications and broadening configurability and flexibility. And do that while still establishing that "Success is Assured" for the entire ranges of values that cannot be known at design time.

Chapter 14

Monday Morning, Week 3, UA Headquarters

Getting to the purpose of the morning's meeting, Nathan said, "I am looking forward to seeing the Trade-Off Charts that we get from the Maps we have created. I am very curious to see the feasible design space. So what do you have for us, Alex?"

"Well, before I show you the charts that I have prepared, let me explain where we are and what I expect to happen today. I will be very surprised if the charts that I show you today are actually the charts that you need to see to make the trade-off decisions. I am not an expert in your space, so I don't know what you need to see or how you need to go about analyzing the design space. And I couldn't ask you, because what you need to see will depend on the actual shape of the design space; you will need to see the design space, and based on that, figure out what more you need to see.

"What we've done so far with the Causal Maps is capture *what* Decisions you need to make and the *structure* of the design space. Based on that, we can now make visible to you the *limits* of that design space and the *sensitivities* between those Decisions. With that visibility, you can then determine *how* you want to go about making those Decisions, which will include what parts of the design space that you need to see and in what form you want to see it.

"In a sense, the Maps we have built are the first layer of reusable knowledge — the decisions to be made and the relations between them. The Trade-Off Charts that end up being useful in *how* you make those decisions form a second layer of reusable knowledge. We'll try to work out what's needed this morning, then I'll generate the charts this afternoon, and we can discuss them tomorrow morning. That's why I asked for both today's and tomorrow's meetings.

"And just to complete the picture of where this is going, there is a third layer of reusable knowledge: you will likely develop a series of different charts that help you progressively narrow the decisions, eliminating the infeasible and less desirable options as you go. That sequence of charts and converging decisions will all be captured in the K-Brief that lays out that larger decision logic. Does that all make sense?"

After a pause, Nathan broke the silence: "Vaguely; but I am guessing it will make a lot more sense once I can see the Trade-Off Charts and K-Briefs that come out of this."

Alex smiled. "Yes, let me say that more simply: I don't know what Charts you need to see — I plan to learn that from you today — but you won't be able to tell me what you need to see until I give you visibility to the design space, which means I need to show you my best guess at some useful Charts."

Laura, growing impatient, acknowledged, "Makes sense; so, let's see it!"

Alex pulled up the first Trade-Off Chart (Figure 14.1). "From our opening discussions, you said that you can freely trade off 'Range' for 'Endurance' at mission time. So, unless you know specifically how much of one or the other you want, you'll need to take a look at both together. So, that's the first chart that I created: 'Range' versus 'Endurance.' That lets you see the design space you have to work with at mission time."

Alex explained, "The shaded region is infeasible. The white area is feasible. The points along the boundary show you how much 'Endurance' you have left if you need that much 'Range.' Or vice versa: if you want a certain 'Endurance' at the target, you can see the furthest 'Range' you could be away. You already knew this trade-off existed, so this isn't too interesting. What is more interesting is what happens to this Chart as we change other design decisions. For example, this was computed assuming that the 'Area Tracked' would be set at five square kilometers. What happens if we choose 1? Or 11? That's what this next chart shows." (Figure 14.2.)

Alex explained, "Each colored band is shading out the infeasible region at that specific 'Area Tracked.' Each boundary is showing you where the trade-off curve is for that 'Area Tracked.' As you track more area, you get less 'Range' or less

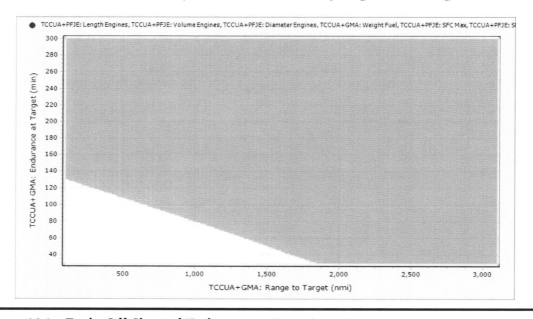

Figure 14.1 Trade-Off Chart of 'Endurance at Target' vs. 'Range to Target'.

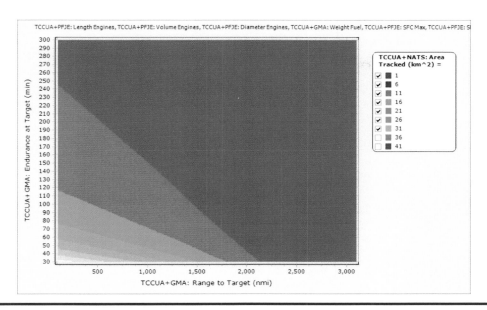

Figure 14.2   Trade-Off Chart of endurance vs. range for different areas tracked.

'Endurance' or both. You can imagine each darker band being in front of the others, coming out of the screen, forming a 3D surface."

Teresa asked, "Why not show that third axis as a Z-axis in a 3D chart?"

"I could have done that. But the 3D perspective can make it hard to see non-linearities. For example, in this chart you can see that the boundaries are initially spread apart but gradually get closer together as 'Area Tracked' grows larger; the effect of 'Area Tracked' on 'Range' and 'Endurance' is non-linear.

"Further, with inequalities, where you need to shade out the infeasible regions, that can be very hard to see in 3D, as you start to see the various edges of the chart showing through the translucent regions, confusing what is what.

"It doesn't take long for our brains to get used to reading contour lines like this; even without inequalities, I consistently find it easier to see what's going on, particularly when you need to compare multiple charts side by side to look at the fourth, fifth, and sixth dimensions. You need things to line up consistently. You can't do that when you have rotated 3D images."

Teresa nodded, "Okay, I see that."

Alex continued, "So, by letting you visualize that 3D surface, you are effectively seeing an infinite number of possible designs — every possible 'Area Tracked' from 1 square kilometer to 41. But there are many other decisions that we can change that will impact 'Range' and 'Endurance.' I generated several of these Charts, each with different combinations of different values for each of the other decisions you might want to trade off. This one assumes a 'Size of Target' of 0.4 square meter, a 'Delta Temperature of Target' of 6 kelvin, and an 'Altitude' of 60,000 ft. It also assumes a 250 ft 'Takeoff Distance' with 0 ft 'Takeoff Clearance' and 50 ft 'Takeoff Altitude.'

"I can show you the same chart but with 'Altitude' at 80,000 ft. Or I can show you this chart that shows how the 'Endurance' versus 'Range' trade-off moves with

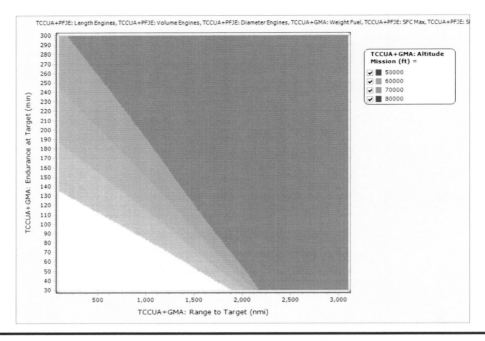

Figure 14.3   Trade-Off Chart of endurance vs. range for different altitudes.

'Altitude.'" (Figure 14.3.) "This chart fixes 'Area Tracked' at one square kilometer. Note the bands are evenly spaced, so the effect of 'Altitude' is fairly linear. In contrast, the effect of 'Area Tracked' in the previous chart was clearly non-linear."

"Another key trade-off is with 'Takeoff Distance.' Rather than the 250 ft assumed in the two previous charts, this shows how the 'Endurance' versus 'Range' trade-off moves as you vary 'Takeoff Distance.' Again, non-linearly." (Figure 14.4.)

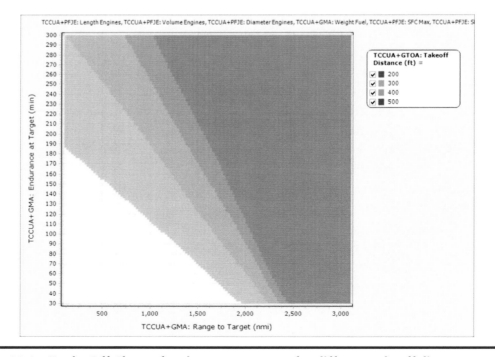

Figure 14.4   Trade-Off Chart of endurance vs. range for different takeoff distances.

Squinting at the charts as if trying to see something that can't quite be made out, Laura suggested, "I think these charts aren't really showing me what I need to see. I'm not really trying to optimize a single airplane. I'm trying to optimize the tracking capability of an aircraft carrier. That capability could be provided with one large aircraft or with ten much smaller aircraft. I don't think this model will really let me see that. Or would it?"

"Ah, interesting point." Alex quickly brought back up the Navy Map and made a few changes. "We could extend the model to have a multiplier that is the number of aircraft on the carrier. We would then multiply both 'Area Tracked' and 'Cost' by that."

"Good," Laura said with a hint of doubt. "But the bigger trade-off is probably the consumed footprint on the aircraft carrier, which for these purposes can be taken as the full rectangular area that the aircraft covers; in other words, its wing span times its length."

"Okay, I have added that." (Figure 14.5.)

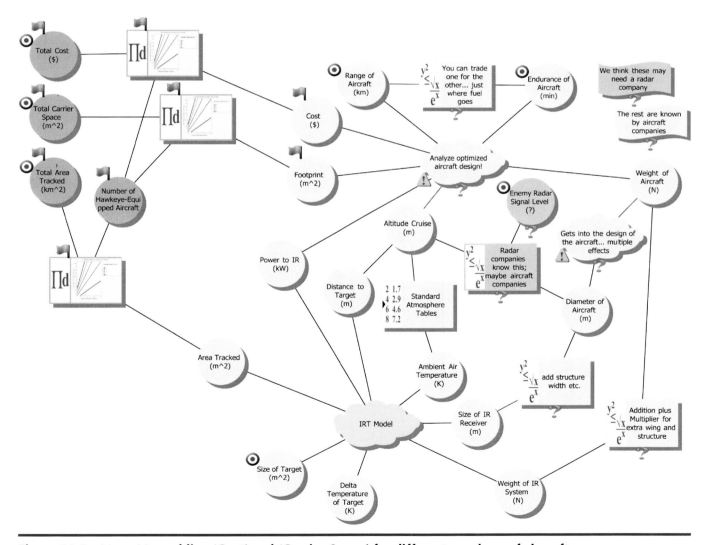

Figure 14.5   Navy Map adding 'Cost' and 'Carrier Space' for different numbers of aircraft.

With more than a hint of skepticism, Laura asked, "So, with that, would we be able to optimize the number of aircraft and the size of those aircraft?"

Alex confirmed, "Yes, for a particular 'Area Tracked.'"

"Oh… Hmmm…" Laura's furrowed brow had returned, but now she was squinting at her own notes. "Can we do that more generically somehow? I ask because there's not really one answer for 'Area Tracked.' That's really a decision we want to delay until we are deploying the aircraft carrier."

Joe answered, "We run into such scenarios in our own system designs. We often deal with it by optimizing the ratios. If I understand you properly, what you really care about is minimizing the ratio of the 'Footprint' consumed on the aircraft carrier to the 'Area Tracked.'"

"Yes! And minimizing the ratio of 'Cost' to 'Area Tracked.'"

Alex added in those ratios (Figure 14.6). "Okay, I have that. Yeah, that simplifies things. I can delete the other stuff I was adding, but I'll leave it for now. The ratios show you what you care about for the aircraft design, in a way that

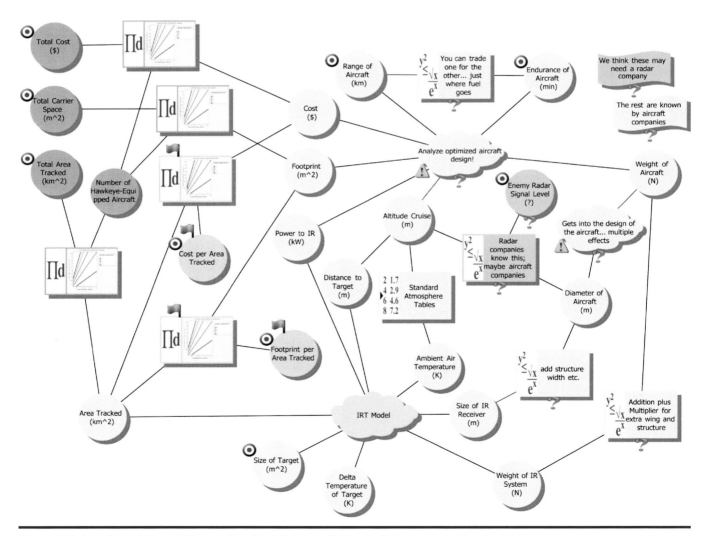

Figure 14.6 Navy Map adding ratios for 'Cost' and 'Footprint per Area Tracked.'

is generically true for all different mission scenarios. No matter how much 'Area Tracked' you need, and no matter how much space you have on the aircraft carrier, you want an aircraft that minimizes the ratio of 'Footprint' to 'Area Tracked' and 'Cost' to 'Area Tracked.' Nice!"

"Hmmm…" With more of a positive tone, Laura asked, "If those are trade-offs against each other — which they could easily be, looking at this Map — then being able to see that trade-off would be hugely helpful in deciding where we want to be in the design space. I am having trouble visualizing what that might look like, but logically it makes sense. Can we look at that tomorrow, Alex?"

"Yes, I can create that. What else would you like to see tomorrow?"

Can you imagine what those Trade-Off Charts will look like? No? That's okay! Generally speaking, for complex systems, nobody can. We simply have to compute them. But once they are made visible to us, we can easily interpret them. And we can easily discuss where in those charts we would prefer our design to be. That is the power of Trade-Off Charts.

A corollary to this is that for similarly complex systems, it is unfair to ask your customer where they want to be on the trade-offs unless you make the Trade-Off Charts visible. Without doing so, you are asking them for the impossible. You might get an answer, but it certainly won't be reliably accurate. They realistically may not even be able to visualize roughly what those trade-offs might look like, let alone provide proper guidance on where they'd want to be on them.

Hence why, as the complexity of our products rise, our requirements and specifications seem to become more and more unreliable and ever more fluid later in the process. Visualizing the nature of the trade-offs becomes more difficult, and thus choosing targets for your requirements that are on those non-visible trade-off curves becomes highly unlikely, until later in the process, when they start to become visible.

Chapter 15

Tuesday Morning, Week 3, UA Headquarters

Alex opened, "So, I added the two ratios we discussed yesterday: 'Footprint per Area Tracked' and 'Cost per Area Tracked.' And since those seemed to be the key customer interests that Laura wanted to trade off, I put those two on the X- and Y-axes of this Chart." (Figure 15.1.) "But of course, those will trade off with numerous other things in this multi-dimensional design space. Since we have a major knowledge gap around Radar Signal Level and Survivability and all, I decided to put 'Altitude' on the third axis — on the contour lines shown in the legend here."

Laura's furrowed brow had returned. "So, what is this telling me is the best design?"

Alex answered, "The white area is the feasible design space if you want 'Altitude' to be 80,000 ft. There are an infinite number of design points in that white design space that all work. The feasible design space at 65,000 ft is anything not shaded out by green, and at 50,000 ft, anything not shaded out by purple. What's 'best' depends on what trade-off you prefer to make between Footprint and Cost. Note that the lowest Footprint is not the lowest Cost, but is pretty close. The lowest Cost is actually quite a bit above the lowest Footprint.' But you could reasonably choose any point on the curve between those as being 'the best,' depending on your priorities. (That is sometimes called the 'Pareto front')." (Figure 15.2.)

Mark further observed, "So, if we want to fly at a higher altitude for better survivability, then we will need a larger footprint and more cost to track the same area. That makes sense. And as we go higher, it appears that impact is increasing non-linearly, since the Pareto fronts of each curve are spaced further apart the higher we go. Am I reading that right?"

Alex responded, "Exactly. But that is just one of the other dimensions — one of the other key trade-offs. In this chart, I am showing the case for 300 ft 'Takeoff Distance,' 2100 nautical mile. 'Range to Target,' Mach 0.7 cruise speed,

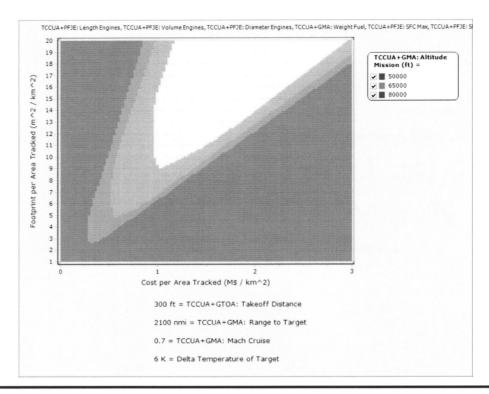

Figure 15.1   Trade-Off Chart of 'Footprint per Area Tracked' vs. 'Cost per Area Tracked.'

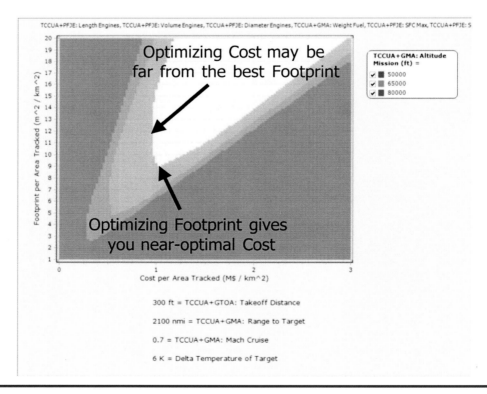

Figure 15.2   The Pareto Front is the boundary from the two different optimal points.

Figure 15.3 Same Trade-Off Chart for different takeoff distances.

and 6 kelvin 'Delta Temperature of Target.' I have similar charts for different values of each of those so that you can start to get a feel for the impact of each of those decisions on both the Cost and Footprint Ratios. For example, if we look at 'Takeoff Distance' at 200, 300, and 400 ft, we get these three charts." (Figure 15.3.)

Leaning toward the screen, Laura interpreted what she was seeing: "Okay, so going from 300 to 400 ft 'Takeoff Distance' reduces Footprint Ratio and Cost Ratio, but only a little bit — probably not enough to justify needing assisted takeoffs. But reducing it to 200 ft to give us some extra padding increases the ratios a lot — and more so for the higher altitudes. So, if 50,000 ft is okay, then that might be a good trade-off. But if we want to be at 80,000 ft, then that is not a good trade-off! Am I reading that right?"

Alex responded, "Exactly. So, there's a non-linearity with 'Takeoff Distance'; and that non-linearity gets worse non-linearly with 'Altitude.'"

Laura grimaced a bit at that comment. "Uh, I guess I see that. So, if it is non-linear, I guess we might be able to get a little padding cheap — maybe 280 ft? How do we figure that out? I guess we just keep generating charts like this at different values?"

Alex answered, "Well, you could. But we'd rather just look at the whole infinite set. That's the challenge with complex problems that have more than three trade-offs: our brains have trouble seeing more than three dimensions at once."

Given most design problems involve far more than three trade-offs, our Success Assured® software tools for Set-Based Design include advanced ways to visualize these multi-dimensional trade-off spaces. However, we wanted to keep this book focused on what you can do using standard charting tools such as Excel. So, Alex will stick to showing three dimensions at a time here. But we didn't want you to walk away not knowing there are better ways to visualize multi-dimensional trade-offs. However, the techniques Alex uses to get down to just three dimensions at a time are useful even when you have the more advanced visualizations, so we would want to teach the following anyway.

Alex continued, "So, to let us see the effect of these other decisions, I took advantage of the fact that optimizing the Footprint Ratio tends to be close to

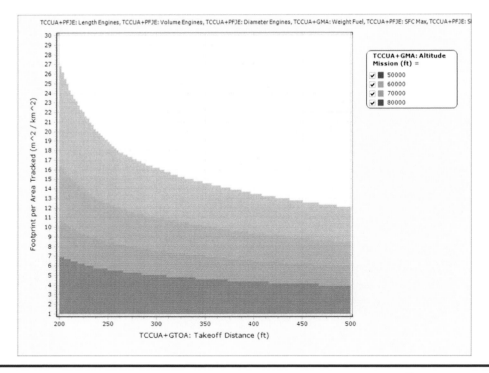

Figure 15.4   Trade-Off Chart of 'Footprint per Area Tracked' vs. 'Takeoff Distance.'

optimal on the 'Cost Ratio.' So, in the following charts, I just show Footprint Ratio, knowing that if you pick a low Footprint Ratio, you will tend to have a low Cost Ratio. In this first chart, I just show Footprint Ratio versus 'Takeoff Distance,' to make it easy to see that trade-off." (Figure 15.4.)

After a pause to orient herself, Laura observed, "So, the curve starts getting a lot steeper at about 260 ft, depending a bit on 'Altitude.' But dropping to 275 ft or so would likely be a good trade-off as it makes it far more reliable to not need assisted takeoff."

Alex continued, "We can look similarly at each of the other key trade-off decisions that you need to make. For example, Range and Endurance were key customer interests. The first chart I showed was for a Range of 2100 nautical miles and Endurance of 180 minutes. If I vary the Range down to 1100 or up to 3100 nautical miles, then you get these charts." (Figure 15.5.)

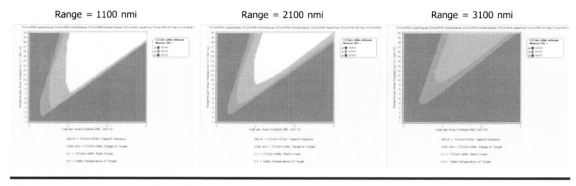

Figure 15.5   Same Trade-Off Chart for different ranges.

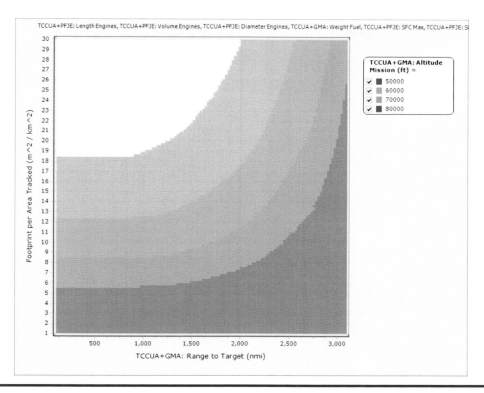

Figure 15.6   Trade-Off Chart of 'Footprint per Area Tracked' vs. 'Range to Target.'

Mark quickly interpreted what he saw, "So, 'Range' has a bigger effect on the Footprint Ratio than on the Cost Ratio. And it increases non-linearly with 'Range' – and again more so with the higher 'Altitude.' So, Alex, I assume you have a chart showing the Footprint Ratio versus 'Range' to allow us to see that better?"

"Yes, I do." (Figure 15.6.)

"And I have a similar chart of the Footprint Ratio versus 'Endurance at Target'." (Figure 15.7.)

After a pause to absorb these two charts, Mark pondered, "It is interesting that below a certain point they both seem to go flat."

Teresa answered, "Well, the other requirements of the aircraft are going to give you a certain minimum Footprint that will have a certain volume available for fuel. Below that, further reducing desired 'Range' or 'Endurance' doesn't really change the design of the aircraft. So that makes sense."

Alex continued, "Another key customer interest decision in our Causal Maps was the cruise speed — how fast could you get to the target. The prior charts had all assumed a 'Mach Cruise' of 0.7. However, when I looked at a 'Mach Cruise' of 0.5, both the Footprint Ratio and Cost Ratio got worse." (Figure 15.8.)

"So, since faster is better, it seems there's no point in looking further; it is win–win to go faster. Well, unless perhaps there's a peak or valley; perhaps the best case is in between. Looking at the Footprint Ratio versus 'Mach Cruise,' you can see that the peak happens to be at roughly 0.7. Above that it starts to get worse again. It is just above 0.7 at high 'Altitude' and just below 0.7 at lower 'Altitude'." (Figure 15.9.)

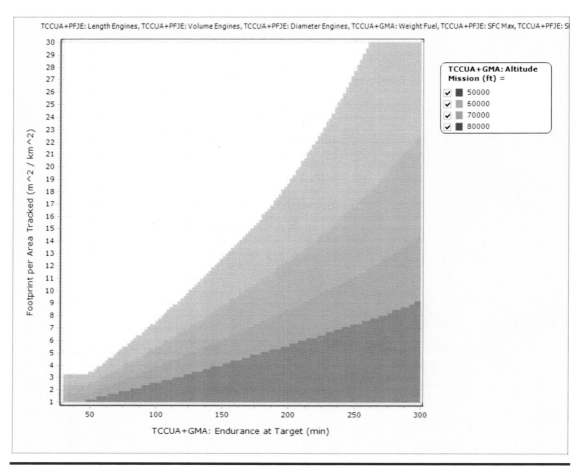

Figure 15.7   Trade-Off Chart of 'Footprint per Area Tracked' vs. 'Endurance at Target.'

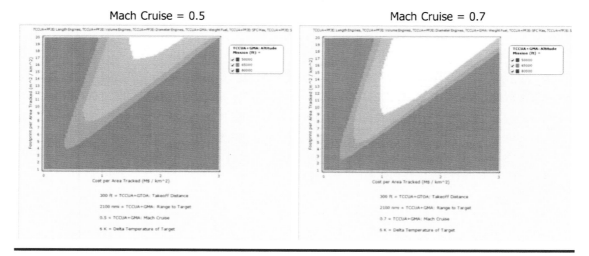

Figure 15.8   Same Trade-Off Chart for different 'Mach Cruise' values.

Teresa concurred, "Yeah, that's not too surprising. The actual velocity of a particular Mach number changes with altitude (it is relative to the speed of sound, which changes with altitude due to changes in temperature). That's why we use Mach speeds in aerospace design discussions, because the behaviors at Mach speeds tend to be less affected by altitude. So that makes sense."

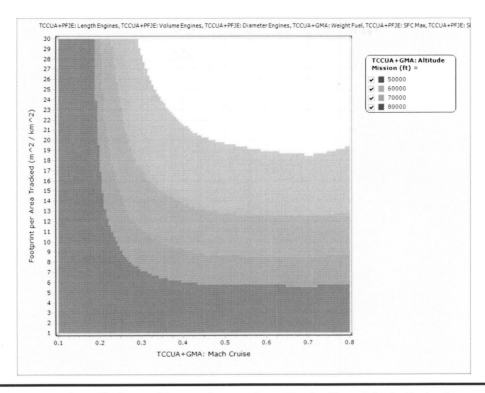

Figure 15.9 Trade-Off Chart of 'Footprint per Area Tracked' vs. 'Mach Cruise.'

Joe stepped in with an IRT concern: "A key knowledge gap from our side in the design of the Hawkeye system for this is the smallest 'Delta Temperature' of the target versus its surroundings that you want to be able to track. That combined with the altitude will drive the required power. So, Alex, can you bring up that trade-off?"

"Sure. Here's the Footprint Ratio versus Cost Ratio charts at 'Delta T' of 2, 6, and 10 kelvin. The difference is small, going from 10 to 6 kelvin, but the difference gets a lot bigger going down to 2 kelvin at the higher altitudes. There is just a small effect if 'Altitude' is only 50,000 ft." (Figure 15.10.)

"To see that effect more clearly, I also charted the Footprint Ratio versus 'Delta T'." (Figure 15.11.)

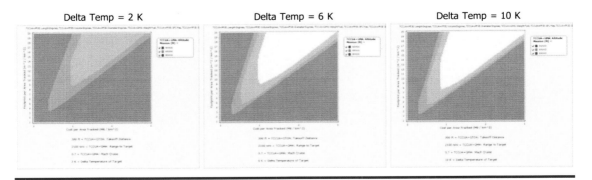

Figure 15.10 Same Trade-Off Chart for different 'Delta Temperature' values.

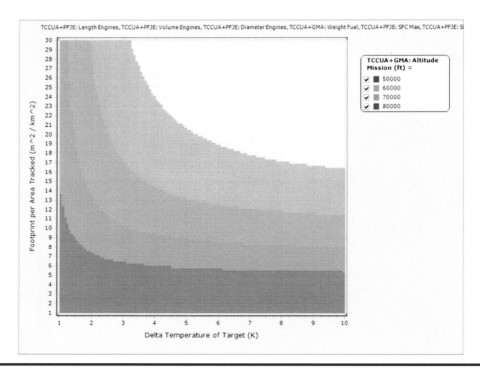

Figure 15.11 Trade-Off Chart of 'Footprint per Area Tracked' vs. 'Delta Temperature.'

Joe explained, "The knee in the curve is at 2 to 3 kelvin for 50,000 ft but up around 6 or 7 kelvin at 80,000 ft. So, that's an important trade-off that the Navy will want to decide on. Is the 'Altitude' and the related survivability more important than the lower 'Delta Temperature' detectability? That will depend heavily on what types of missions you are targeting. Hot desert environments tracking humans? Or colder environments? Or tracking vehicles with hot engines? And how high do you need to be for survivability? We will need to know where you want to be on those curves."

Laura nodded. "Yes, that would be very tough to answer generically; we want it all, so to speak. We couldn't prioritize one versus the other. However, given the visibility of these curves, we may be able to determine spots that are good enough in every aspect to satisfy our target missions."

This is a particularly important point. If you are skimming, you may want to take some time to understand this particular Trade-Off Chart and think through how you would prioritize these key customer interests if the curves were shaped differently.

Asking customers for priorities between their many requirements is largely asking for the impossible, because the answer depends on the specific levels that you can achieve and how those align with the specific combinations of levels they need for specific application scenarios. Without the supplier giving the levels that can be achieved, the customer can't really do that mapping and thus cannot answer the relative priorities.

Of course, the net result is the customer provides the guidance as best they can without that visibility, guessing at the levels of things. When the design later matures and the levels become known, then they can do the mapping to their scenarios, and suddenly they start asking for different priorities or more demanding levels. The supplier then complains about shifting requirements, but in fact it was the supplier's fault for not providing the required visibility to the trade-off curves (containing the specific levels) up front.

Tim and Teresa had been whispering to each other and pointing at the screen for the prior several minutes; as the level of concern on their faces continued to grow, it began to concern the rest until Nathan asked, "Is there something wrong here? You two seem to be concerned with one of the charts that Alex has up."

Tim apologized: "Sorry, Teresa and I went off on a tangent — nothing related to our discussions here. We have been spending countless hours in heated meetings for the last, what...?"

Teresa, shaking her head, filled in an estimate, "I don't know, three months?"

"Oh, it has to have been longer than that; it was the hot topic at our last executive review. We're probably approaching five months. Experts in different areas are each strongly championing very different design approaches for an aircraft not too different from this one but with a very different application. They have formed into three camps, each working their designs, each running into different difficulties, each arguing they need help from the others to accelerate their work, each arguing the others' approaches are clearly inferior to theirs."

Teresa added, "It has been painful and ugly — very inefficient. Some have argued that it is okay, that we are just operating Set-Based and thus it's a good thing. However, after today, after seeing the Set-Based work that we've been doing here, I can firmly assert there's nothing Set-Based about that other effort."

"Agreed!" Tim seconded that and continued, "Anyway, looking at these charts, Teresa and I are pretty sure that all three of those camps are each working designs that are well into the infeasible regions of these charts. In other words, we think the Set-Based analyses done here show that all three *points* that we have spent so much time developing and arguing over the last five months are all fundamentally unworkable."

Teresa, still shaking her head, added, "And we did this Set-Based analysis in a matter of a few hours over the last couple weeks. With all the expertise we had in just one of those extended heated meetings, we could have probably done this in a day."

Tim, imagining how things might have gone had they done that, added, "And if we had been focused on one of these feasible regions, I wonder if those teams would have run into far fewer issues and been able to flesh out their designs much more quickly without running into so many obstacles. Ah, the time wasted."

Scott, similarly imagining the consequences, followed, "And just think, had the team done that, then when the Navy came calling on this Hawkeye project, we'd have been able to do as PFC did with their jet engine model and just pull out our Set-Based models and show them what's possible and what's not — show them the shapes of the trade-offs."

Tim looked over at Teresa and built on that thought: "How many hours do we spend in vague discussions with our customers trying to guide them to parts of the design space that our experts feel more comfortable in, but where we have no clear data, so those conversations take forever? And in the end, we're often making compromises with our customers that get us into infeasible spaces anyway."

With a smile, Alex shrugged. "But then we'd have missed out on all the fun mapping out this aircraft design!"

Teresa chuckled. "Actually, that's a good point; we ourselves have been missing out on a lot of the fun we should have been having mapping out these designs. Exploring the design space based on real data is fun; arguing over opposing expert opinions is no fun at all. We, of course, already generate lots of trade study data to support our arguments, but without the mapping effort giving us a complete picture of what impacts what, there are always major caveats, always key things missing, always critical doubts that leave us guessing where we really are. In what we have done here, we *have* made simplifying assumptions, but we have *bounded* them such that if we know this is the worst-case performance, then we don't fall back into those endless arguments over opinions on those assumptions."

With a smile, Alex piled on: "We would definitely agree on all that — but there is actually even more fun to be had. You are not only getting clarity on what's feasible from the Trade-Off Charts, that underlying Decision Map is giving you clarity on what is connected to what — to what is causing those feasibility limits – to what is driving those sensitivities. So, when the spot you want to be in is in the infeasible space, you can focus your team's innovation on precisely the limits that are getting in your way. And better yet, you can compute exactly how far you'd need those causal decisions to change to move those limits just far enough to hit your target. Very often, that exposes clever alternatives that can do that. That sort of productive, focused brainstorming can be a lot of fun!"

This is very much a real-world situation. Our clients have shown us cases where they were able to demonstrate that a project had been working for months or even years to achieve a target in parts of the design space that were provably infeasible with Set-Based methods. In some cases, they were far outside of the feasible regions; no amount of innovation was going to save that project.

Although no team has time to go back and perform post mortems on failed projects, once they build the Set-Based knowledge for their existing projects, it often takes very little time to simply plug in the numbers for those past projects. And far too often, the answer is that those failed projects had no chance of succeeding due to those first oh-so-critical targeting and approach decisions. But with no way to see the design space and no way to do the analyses except based on a particular design point, they had to perform the whole development process to come to that realization. And even then, they shut down the project not knowing how far they may have been from (or how close they were to) feasible.

Not only is it wasteful of company resources, it is frustrating and demoralizing for the people involved.

Ready to get back on topic (her topic), Laura asked, "Okay, so what's next?"

Joe answered, "You have some knowledge gaps to close. We have established and made visible to you considerable design space for you to work in. But where you want to be in that space depends on how you prioritize the Footprint Ratio versus the Cost Ratio versus the Altitude versus the Delta Temperature versus Range and Endurance. All of those trade-offs are impacted by the Altitude, and that is driven by the Survivability knowledge gap. So, if I were the chief engineer for this system design, I would be setting up a series of Integrating Events to make those decisions, based on when those decisions need to be made, and then focusing my teams on what they needed to learn to make those decisions based on knowledge before those Integrating Events."

After a bit more thought, Joe continued, "While working out that series of Integrating Events, I'd be calling meetings with experts in different areas, presenting the Maps and Charts to them, and asking them, 'What's missing? What other things might be impacted? What other decisions might affect these? Who else should we be talking to?' That is one of the most powerful benefits of this process and these visuals: you can quickly and efficiently engage experts in many different areas in order to identify all the key decisions and all the key knowledge gaps."

"Thanks Joe," Laura responded. "I see your logic for the Integrating Events and the first steps make sense. Would you mind emailing those specific suggestions and I will take a pass at laying out the next steps and timeline."

"Will do," Joe responded.

With that cue, Mark decided to raise a new issue: "Laura, the switch to Footprint Ratio and Cost Ratio has certainly simplified this analysis — it has allowed us to ignore the decisions regarding the number of aircraft versus the size of those aircraft. However, I wonder if there are some important trade-offs there. For example, will the maintenance costs go up having more small aircraft rather than a few larger aircraft? But then, at the same time, the overall reliability of the fleet goes up because having one aircraft down has only a small impact on your mission capability. So, is it better to maximize reliability with more smaller aircraft or minimize maintenance costs with fewer larger aircraft? I think we should have an Integrating Event to make that decision and decide how to factor that into this analysis."

Nathan built on that: "Interesting point, Mark. And that feeds into a different concern that I have been pondering: although we chose when we started all this to keep things simple and just focus this aircraft design on the Hawkeye missions, we have additional uses in mind for this aircraft; it won't be dedicated only to Hawkeye missions. Will those other missions require a minimum-sized aircraft? If so, how do we fold that into this analysis?"

Joe responded, "In my experience, it may be as simple as clarifying the minimum size for those other applications and the size below which you hit a knee in the maintainability curve, and then add to this analysis that minimum as a simple limit on the aircraft size. That will cut off parts of the design space that you are currently seeing, and then you can proceed as you would have otherwise."

Alex added, "Where that won't be sufficient is if those requirements skew the slopes of those curves such that they become trade-offs that have to be considered together. However, you'll know that if you extend your Causal Map with those other factors. If they don't tie into the existing Map anywhere but the minimum aircraft size, then you are good. But if they have trade-offs with other decisions in the Map, then you may need to work those areas together."

Nathan leaned toward Laura. "I see some more Causal Mapping sessions in our near future."

Laura nodded. "We definitely need to pull in both Nick and Maria; they have each been working on requirements for the aircraft for their own objectives. We won't be able to make any decisions until we have their issues incorporated."

Laura looked at Mark. "Mark, are you ready to play Alex's role?"

Or a better question: Are you, our reader, ready to play Alex's role in your next design discussion? Are you ready to take notes in the form of a Causal Map, capturing each measurable decision as a circular shape and each relation or limit between those decisions as rectangular shapes?

The intent of this book is twofold: to get you ready to play that role and to get your leadership asking you to play that role.

Chapter 16

Friday Morning, Week 3

Laura and Mark were already in the meeting room, Causal Map up on the screen, when Maria arrived.

"Hi, Maria," Laura opened. "Thank you for your time today."

Maria cut in: "Before we start, let me warn you that the mission scenario that I have been working on is classified; I won't be able to discuss it with you. I can't really even communicate the full set of requirements; they want me to keep it to just the requirements that you do not already satisfy."

Laura looked over at Mark for verification. "Normally, I'd find that very irritating, but I think this process actually accommodates that pretty well." Mark nodded.

Looking back at Maria, Laura continued, "We have been analyzing what we need to put in the specifications using a Causal Mapping process where the goal is to establish knowledge to eliminate the weak parts of the design space. So, what we're looking to get from you are any key things that are missing from our Map and any parts of the design space that you want to eliminate as unworkable for your mission needs."

"Okay, great; I can do that!" Maria leaned back in her chair and turned toward the screen, taking on the reviewer role.

Laura looked over to Mark. "So, Mark, let's start with the higher-level Map to set the context: our mission needs."

Mark pulled the Map up (Figure 16.1).

"Each circle shape in this Causal Map represents a design decision that needs to be made. The ones marked with red target symbols are the customer interests – the things we want more or less of. We want to put up some number of IRT's new Hawkeye IR Multi-Tracker devices to track an area of some size. On the left, you see the customer interest decisions 'Total Area Tracked,' 'Total Carrier Space,' and 'Total Cost,' which all depend on the 'Number of Hawkeye-Equipped Aircraft' that we put on the carrier. However, since that can vary per mission, we are not using those purple shapes; instead we are optimizing the two ratios in green: 'Cost per Area Tracked' and 'Footprint per Area Tracked.'

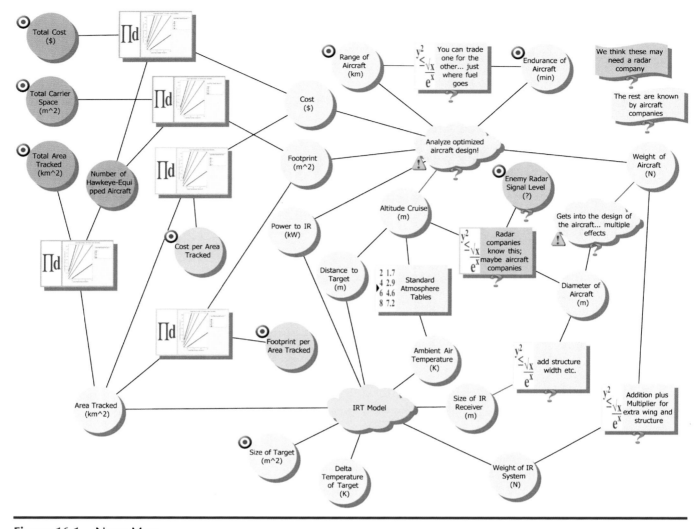

Figure 16.1   Navy Map.

"The other customer interest decisions that will impact those ratios are the 'Range' and 'Endurance' we want (up top), the 'Altitude Cruise' we want in order to minimize the 'Enemy Radar Signal Level' for survivability, and the 'Size of Target' we want to track and its 'Delta Temperature' versus the environment.

"The cloud shape indicates numerous other shapes go there; in this case, they indicate another Causal Map goes there. For example, near the bottom is a cloud labeled 'IRT Model' — they have that part modeled — we can just assume that if we give the ranges of the decisions connected to that cloud ('Area Tracked,' 'Size of Target,' 'Delta Temperature of Target,' 'Weight of IR System,' 'Size of IR Receiver,' 'Ambient Air Temperature,' and 'Distance to Target' [which is essentially 'Altitude']), then they can tell us what subsets of those ranges will work and how much 'Power to IR' they will need.

"'Power to IR,' 'Size of IR Receiver,' and 'Weight of IR System' are together the payload burden on the aircraft design, which is modeled by these two other clouds. We have a separate Causal Map that we worked out with the UA engineers to fill in what's in those two clouds.

"Before we look at that Aircraft Sub-Map, which is probably the one of greater interest to you, do you have any questions on this Causal Map?"

Maria paused for a moment to ponder that question. "Wow, I feel like that was more than I could possibly take in so fast; however, it all seems to make sense and I don't have any questions right now. But I reserve the right to come back and ask some questions after I see the Aircraft Map."

"Yeah," Laura chuckled, "we'll be happy to do so. This next Map will be a little more daunting; we should probably have broken it up into pieces to let you grow into it step by step, but we didn't do that, sorry. We'll see how this goes. Mark, zoom in on the mission stages across the top."

Mark switched to the Aircraft Map and zoomed in on the upper right (Figure 16.2).

Starting from the top, Laura explained, "The scroll shapes are just general statements, not specifically Decisions (the circles) or the relationships between the Decisions (the rectangles) — essentially anything else. Here we have listed the series of mission segments: 'Warm Up and Takeoff,' then 'Climb' to cruise altitude, then 'Cruise To' target, then 'Loiter at Target' (which in this case may be the bulk of the mission), then generically 'Combat' (for us that would just be an avoidance maneuver and sprint to escape), then 'Cruise Back,' and finally 'Loiter to Land,' and 'Land' back on the carrier. Any questions on that?"

Maria responded, "I assume based on some of what I can see here that we can specify different values for the Decisions that correspond to each of those segments; for example, 'Endurance at Target' could be set to zero, effectively eliminating the 'Loiter at Target' segment — true?"

"Yes, exactly."

"Okay," Maria continued, "then I don't have any issue with the sequence of mission segments, but I need to see what ranges of values you are accommodating for the associated Decisions."

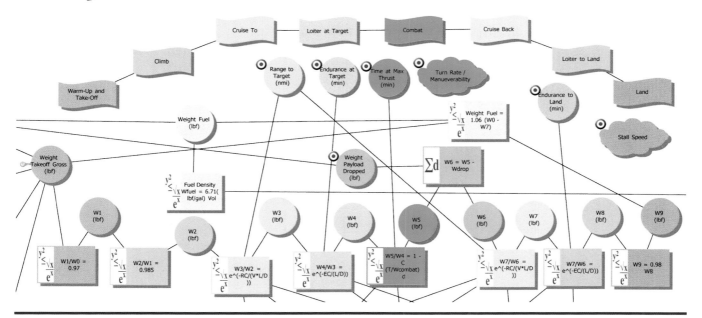

Figure 16.2   Segment weights and targets colorized by mission segment.

"Yes, and watch for things you would want to specify that we don't have a Decision for; those would be the most problematic."

Laura continued, "Running along the bottom is a Relation shape for each mission segment that computes the fuel consumed, and thus the drop in weight, for each mission segment. The Decision shapes in between each is the Weight at the end of each segment, and thus at the start of each following segment. Make sense?"

Maria responded, "Yes, and I see those calculations are being driven by some of the customer interest decisions you have under each mission segment that they correspond to. For example, you have 'Range' to drive fuel consumption for 'Cruise To' and 'Cruise Back,' and you have 'Endurance at Target' for the first Loiter segment and 'Endurance to Land' for the second Loiter segment."

"Exactly."

"But two of these customer interests are cloud shapes instead of circles, and they aren't connected to anything. What does that mean?"

Laura looked back at the Map, "Oh, for our mission, we did not anticipate maneuverability would be an issue, so we didn't bother modeling those details here. But I think the UA guys were comfortable they knew how to map that out. The same for 'Stall Speed'; given the assisted landing, we didn't anticipate 'Stall Speed' to be worth modeling for now."

"For the mission I am working on, Stall Speed will be critical. We would like that as low as possible."

"Okay," Laura responded, "we will work with the UA engineers to get that connected in. Mark, scroll to the left so we can look at the Takeoff Gross and Empty Weights." (Figure 16.3.)

Laura continued, "'Weight Takeoff Gross' is near the center with the key symbol on it; that Decision goes into lots of other things. If you look at the Relation above and to the left of that, you see it is the sum of the Empty Weight, the Payload Weight that is not dropped at the target, the Payload Weight that is dropped at the target, and the Weight of the Fuel, which is summed up from the weight drops in each mission segment.

"Of course, 'Weight Takeoff Gross' is the weight at the start of the first mission segment, and the heavier it is, the more fuel that will be consumed, and thus that's a big circular calculation. Worse, there's another circular calculation to the left of that: 'Weight Empty' is computed from 'Weight Takeoff Gross'; you need to size the structure to hold the maximum weight. That calculation is a rule of thumb derived from historical data, which is dependent on 'Aircraft Type' (e.g., a bomber vs. a fighter vs. a transport).

"There's actually several similar rules of thumb derived from historical data dependent on 'Aircraft Type'; but before we go there, any questions on 'Weight Empty' or 'Weight Takeoff Gross'?"

Maria shook her head, "No, makes sense. I knew there was some sort of circular calculations that forced the aircraft designers to make estimates and have to repeat that loop until they could get it to balance. I wasn't aware of all that was

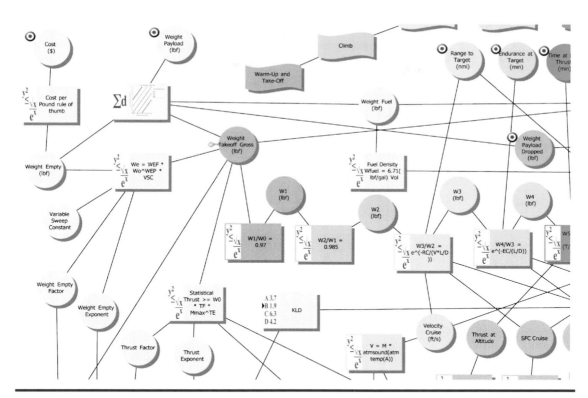

Figure 16.3   Weight sums and cost as function of 'Weight Empty.'

involved before; this makes it clear. In a sense, the whole mission is rolled up into that."

Laura continued, "Mark, scroll down to show everything hanging off 'Aircraft Type'." (Figure 16.4.)

"So, you can see 'Aircraft Type' lower center with the key on it. To the left of it, you see two Lookup Tables that give you the 'Weight Empty Factor' and the 'Weight Empty Exponent' that go into the calculation up top of 'Weight Empty' from 'Weight Takeoff Gross' that we were looking at before.

"Similarly, coming off the top of 'Aircraft Type,' you see two more Lookup Tables giving you 'Thrust Factor' and 'Thrust Exponent' that go into a calculation of how much 'Thrust' you need based on 'Weight Takeoff Gross' and the 'Mach Max' you want to be able to fly at. 'Thrust' is off the edge of the screen; we'll look at the jet engine decisions in a minute.

"And then again we have two pairs of lookup tables coming off the bottom of 'Aircraft Type' that drive similar historical rules of thumb on the 'Aspect Ratio' and 'Length' of the aircraft — together limiting the geometry of the aircraft to shapes appropriate for that type of aircraft. Any questions on that?"

"Hmmm." Maria was a little uncomfortable with that. "I know the aircraft designers use such historical rules of thumb to make their estimates, and that always bothered me a little as it would seem to potentially cut off innovation — only allowing designs that have been done in the past. Often, we end up with disagreements among engineers as to what degree a design is straying away from those estimates as we try to optimize other things. This would seem

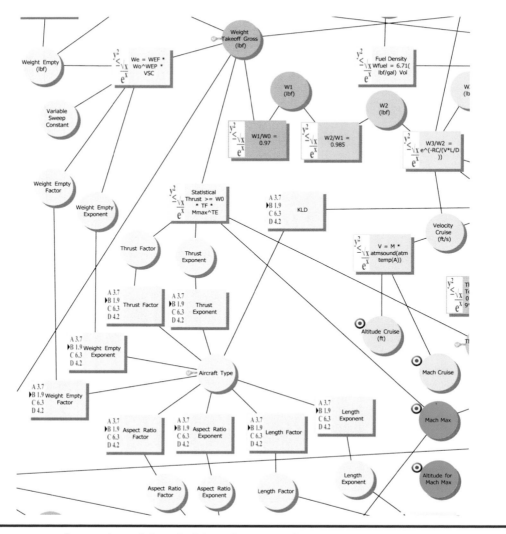

Figure 16.4   Various rules-of-thumb driven by 'Aircraft Type.'

to be locking those things in such that you can't stray — such that you can't optimize."

Mark responded, "Great point; but actually, it is exactly the opposite. Because we are working Set-Based here, the knowledge we are capturing can explicitly represent not just the estimated point but rather the range of values that would be achievable based on past experience. So, for example, the 'Thrust' calculation is not telling you the precise thrust you must have for that 'Aircraft Type,' 'Mach Max,' and 'Weight Takeoff Gross,' but rather the range of thrusts that would be acceptable. So, it allows you to see the larger design space. And it lets the experts capture more than an estimated starting point but rather the full set of acceptable values. The knowledge being captured here is quite rich."

Laura added, "But you may be right that to achieve some performance goal we may need to push beyond the sets that the engineers were comfortable with. In that case, we will need to dig deeper into what the real constraints are — perhaps build up a deeper map that replaces these historical numbers. If, on the other hand, having all these sets visible to us allows us to find a solution in the ranges proven reasonable by past design work, then we can potentially avoid a lot of unnecessary analysis time and cost."

Maria replied, "So, you're not just going to show me these Maps of how things are related, you are going to show me the ranges of values that will work? The sets of designs that will work?"

Knowing she had Trade-Off Charts to show off shortly, Laura smiled, "Yes, exactly."

"Okay, I'm not sure what that looks like, but I am looking forward to seeing it."

"We'll get to that shortly." Laura continued, "Mark, scroll to the lower-left corner so that we can see the takeoff calculations." (Figure 16.5.)

"So, you see the three customer interests regarding Takeoff in the lower left: 'Takeoff Altitude,' 'Takeoff Distance,' and 'Takeoff Clearance.' Ideally, we'd like to be able to takeoff from the aircraft carrier unassisted, allowing us to get a number of these up much more quickly. And that calculation is driven by the 'Takeoff Gross Weight' and will be the key driver of how much thrust is needed — probably much more so than the 'Mach Max' for this application.

"That calculation is heavily dependent on the coefficient of lift, which is related through another lift coefficient to the Sweep angle of the wings, which is in turn also related to the 'Mach Max'. So, yet another set of circular equations.

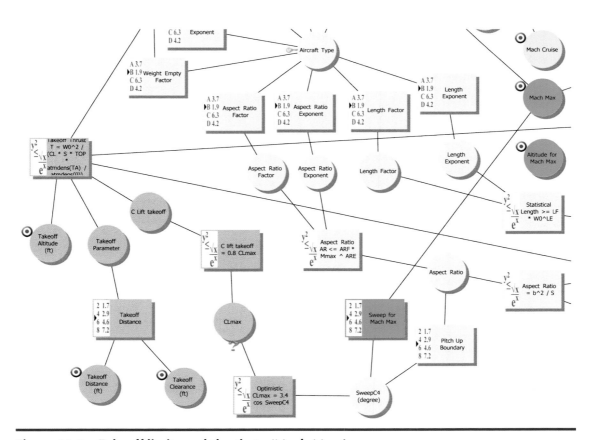

Figure 16.5   Takeoff limits and the tie to 'Mach Max.'

It is also related to the limit on Pitch-Up, which is related to the Aspect Ratio, which as we mentioned before is limited by 'Aircraft Type.'"

Maria smiled and interjected, "So, more circular calculations."

Laura asked, "So, any questions on this? Now that we've taken off, so to speak, and started getting into the mission calculations, do you see anything missing so far from your mission perspective?"

Maria thought for a bit, "No. At this point, if you can take off for your mission, you can take off for ours, given that we can input our Payload Weight."

"Great. Mark, scroll to the lower right so that we can see the geometric calculations — the lengths, diameters, volumes, areas, and aspect ratios." (Figure 16.6.)

"Lower left you see 'Wing Span' and the 'Sref' reference wing area. Those are related to the 'Aspect Ratio' we looked at before and the 'Wetted Area' and 'Wetted Aspect Ratio' you see to the right. The lengths and diameters are driven by the payload dimensions and go into the volume calculations to make sure you can hold all that fuel you're calculating you need. So, more circular calculations.

"Anything missing for your mission needs?"

"No," responded Maria after a long pause to think. "We may need to specify how big of a payload we want to introduce, but other than that I do not expect us to impact anything geometrically."

"Okay," Laura continued. "Ultimately, all of that has to be held up in the air by the 'Thrust' from the engines. The engine model is visually separated by the darker gray colored shapes. Mark, scroll up and left a bit to show the engine model and how it drives into the mission segment calculations." (Figure 16.7.)

Laura continued, "The length and diameter of the engines connect to the aircraft dimensions we were just looking at and the performance of the engine. The effective 'Thrust' of the engine will vary with the 'Altitude' at which we want to

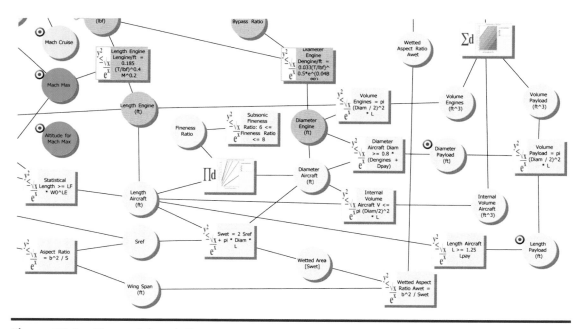

Figure 16.6   Geometric relations.

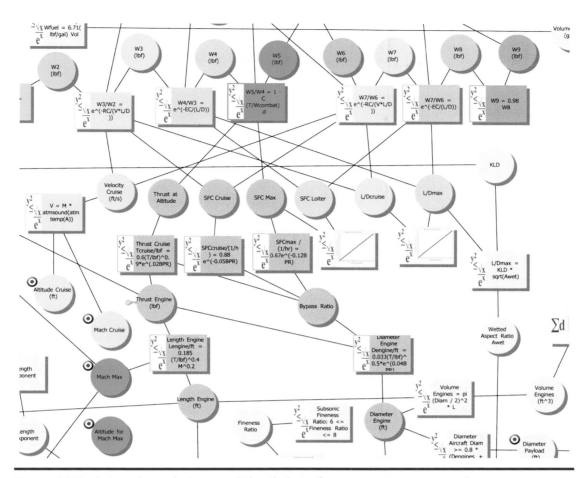

Figure 16.7   Jet engine relations and the tie into the segment range equations.

fly the mission. The engine model gives us the 'SFC,' the specific fuel consumption, that drives the range and endurance calculations for each of the mission segments.

"So, that's the whole Aircraft Map. Besides the 'Stall Speed', are there any other decisions, particularly customer interest decisions, that your mission needs to drive in order to be confident it will succeed?" (Figure 16.8, which is also reproduced across two pages at the end of the book.)

Maria responded, "No, I think everything is there — or will be with the addition of Stall Speed. However, the key thing missing in all this is what values you will be choosing for each of those decisions. That's what I need to understand."

"Great, that's coming next. We haven't chosen specific values for these decisions, but we have made visible the feasible design space for all of them. So, you can look at the design space that remains and determine whether some portion of that design space will work for your mission needs. Or, I guess I should say, you can let us know what portions of the design space do not work for your mission needs, and we'll eliminate those portions as 'weak.' (I am still getting used to this reverse thinking of 'eliminating the weak' rather than 'picking the best.')"

On that, Mark interjected, "If Alex were here, I suspect he'd be warning that the Trade-Off Charts we're about to show are probably not the ones that you will

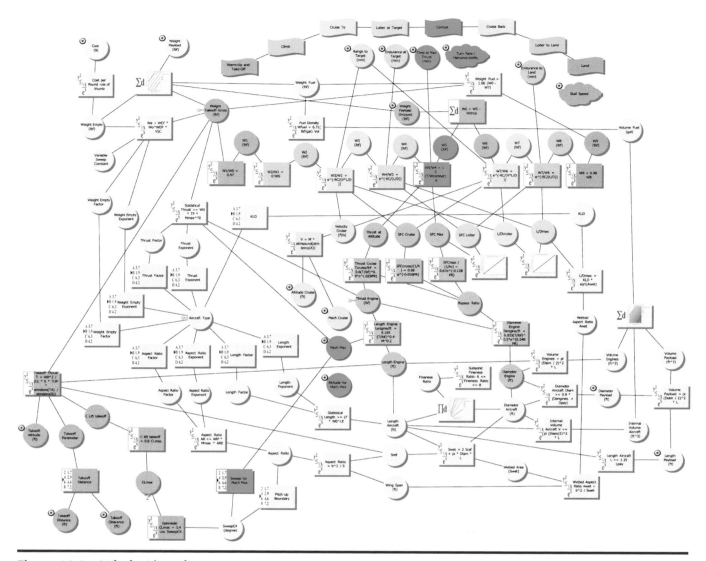

Figure 16.8   Whole Aircraft Map (two-page version at the back of the book).

want to see. But they will give you a better feel for the design space such that you'll know what Trade-Off Charts that you want to see instead."

"Good point. So to start, Mark, pull up the chart of 'Cost per Area Tracked' versus 'Footprint per Area Tracked' with the two different optimal points marked." (Figure 16.9.)

"So, for the preferable Altitude of 80,000 ft, the red area is shading out what's infeasible. The white area that remains is the feasible design space. Of all those possible point designs, the one that delivers the best 'Footprint per Area Tracked' is the bottom-most point in the white. The one that delivers the best 'Cost per Area Tracked' is the left-most point in the white. Note that the best footprint is fairly close to the lowest Cost; in contrast, the best Cost might be 30% larger in footprint.

"If we instead look at an 'Altitude' of 65,000 ft, then we are looking at the green shaded region; pretend the red that is showing is not there. Once again, the

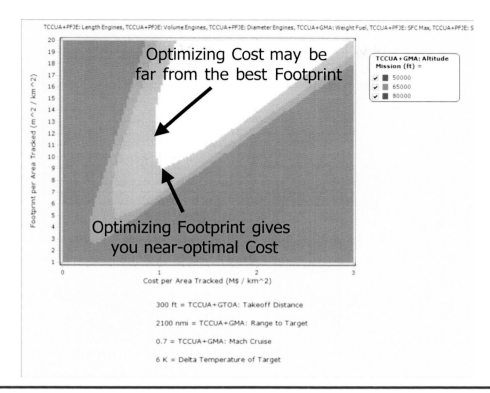

Figure 16.9 Trade-Off Chart of 'Footprint per Area Tracked' vs. 'Cost per Area Tracked.'

best Footprint will be close to optimal Cost, whereas the best cost is quite a bit above the optimal Footprint. Based on that pattern being consistent throughout the design space, most of the rest of the Charts just look at 'Footprint per Area Tracked' knowing that the optimal point there will be near-optimal in cost. But once we narrow to near there, we can then go back to this chart and pick any point between best 'Cost per' and best 'Footprint per.' Does that make sense?"

Maria nodded slowly, "Yeah, I think so; let me see some of those other Charts before I commit to that."

"Sure." Laura continued, "But first, note that this Chart is just one slice through the multi-dimensional design space. For example, this slice is assuming that the 'Takeoff Distance' is 300 ft. What if we allowed it to be 400 ft? Or restricted it to be 200 ft? Mark, can you bring up those three charts side by side?" (Figure 16.10)

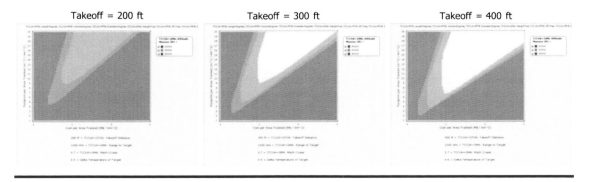

Figure 16.10 Same Trade-Off Chart for different takeoff distances.

Laura continued, "Notice that going from 400 to 300 ft moves the feasible design space at 80,000 ft altitude up a few dollars in cost. But going from 300 to 200 ft moves it up a lot, completely off the top of the chart. So, there is a non-linearity in the effect of 'Takeoff Distance.' But we might need to generate a lot of these charts, sliced this way, to find where the knee in that curve is — to find where we want to be on that curve.

"So instead, we just slice the multi-dimensional design space a different way to make that far quicker and easier to see. We put the decision of concern, the 'Takeoff Distance,' on the X-axis; that will make it easy to see its effect on the 'Footprint per Area Tracked' on the Y-axis." (Figure 16.11.)

"Oh, now I get it!" Maria replied. "You can see the Footprint and therefore the Cost start rising steeply below about 260 ft 'Takeoff Distance,' so you can eliminate that part of the design space as weak but at the same time not prematurely decide on 270 ft; as you optimize other things you may end up needing just a little more space and thus be willing to sacrifice a little 'Takeoff Distance' to get it. Nice."

"Well said," Mark replied.

Maria continued, "So, what is the impact of speed on Footprint?"

"Let me bring up the chart with 'Mach Cruise' on the X-axis." (Figure 16.12.)

Maria leaned back. "Wow, that's a really sharp knee in the curve at lower speeds."

Mark replied, "True, but keep in mind this is assuming you're flying at that speed for a 'Range' of 1100 nautical mile I'm guessing you don't want to spend that much time going that slowly."

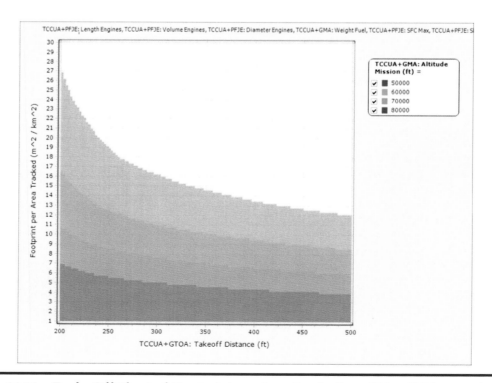

Figure 16.11   Trade-Off Chart of 'Footprint per Area Tracked' vs. 'Takeoff Distance.'

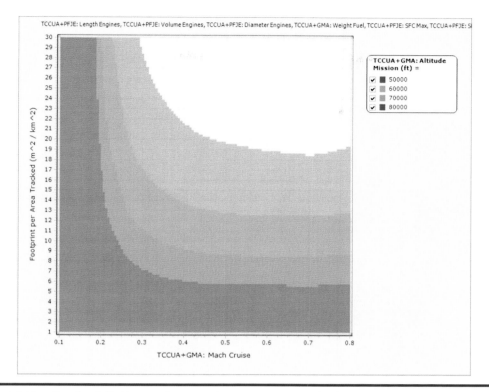

Figure 16.12   Trade-Off Chart of 'Footprint per Area Tracked' vs. 'Mach Cruise.'

Maria nodded. "True. I'll need to talk to my team on how to best represent our mission within your model to gain visibility of what we'd like to see. But this does give me visibility that there is some sensitivity justifying further analysis. I am guessing we can identify the portion of the design space we want to look at, get those Charts from you, and then based on what that looks like decide whether or not we need to add something more to your model."

Maria leaned back in her chair, "Now that I've seen the Trade-Off Charts you're looking at, I would suggest that there are other trade-offs that we will be interested in seeing and may want to influence where you choose to be in the design space. Or, I guess, as you say, we may want to eliminate parts of the design space that we consider weak. In particular, trade-offs against stall speed and trade-offs against size. We want each aircraft as small and capable of slow as possible."

Laura, looking at the Map, asked, "When you say 'small,' do you mean in weight? Or footprint? Or wing span? Or diameter? Or what they call the Reference Area? We have lots of different 'sizes' in this Map."

Maria leaned forward and jotted down some notes on her pad. "That's a good question. We've generally been assuming they're all related, that the whole plane grows or shrinks together. But if you are making trade-off decisions between those things, I guess we need to determine specifically what we are most sensitive to. I don't know that answer, but I think my team knows."

Maria chuckled. "I guess I need to get with my team and map those details out. We won't be able to share that Causal Map with you, but we'll probably need it to help us make the trade-off decisions you will likely be asking us to make to eliminate the weak parts of the design space from our perspective."

After a few moments with a reflective look on her face, Maria continued and concluded. "For years, I have been frustrated that all of our knowledge reuse tools and databases never really provided knowledge we could really use. These Causal Maps and the Trade-Off Charts make it very clear as to why. Unless you can visually see how decisions interact, there is no way to simply transfer data from one set of customer interests to another. This changes everything."

It is often surprising to teams how much more breadth and depth they get from their collaborative discussions when they begin introducing visual models, particularly Causal Maps and Trade-Off Charts. That is as opposed to discussions that ultimately result in statements such as "I agree there may be an issue there; we should probably do some more analysis on that," which are far too vague to be actionable and thus often result in no action at all. In contrast, statements such as "There is a sharp knee in the sensitivity to speed; we need to work out the best way to model this slow-speed mission scenario to clarify where we are on that curve" are highly actionable. The precise details may not be known, but the effect you want to make visible is clear. Further, as shown in these last few chapters, it is an analysis that can be done quickly and made visible to the team to give them deeper understanding of that portion of the design space, allowing them to refocus as necessary on the appropriate next steps. It is that visibility that can greatly accelerate the organizational learning.

Do you understand this Causal Map for the mission design of an aircraft? If so, you understand a substantial portion of Daniel Raymer's fine book on aircraft design, as this model is actually based on that book. For those of you with no education in aerospace engineering, we hope that it serves as evidence of the power of the Causal Map to communicate knowledge across areas of expertise.

Are you an aerospace engineer who has previously studied Raymer's book? Do you now have a deeper understanding than you ever had from the book alone? That's the power of visual modeling! We have spoken to engineers that work for major aircraft companies who have told us that the Causal Maps have given them tremendous insight into the content in Raymer that they had previously missed. More importantly, we have spoken to engineers who have told us that the Causal Maps that they developed from their own knowledge have given them tremendous insight into that knowledge.

Consider doing the same for your own models. Even if you are not going to move the computations out of Matlab or Excel or Simulink or Maple or MDAO or SPICE or wherever you have them, documenting them in the form of a Causal Map can be tremendously valuable to those who need to understand what you are modeling. The improved ability to innovate and push the limits from such visibility cannot be understated.

We hope Part II has made clear that operating Set-Based is not just delaying decisions to give you time to do the learning, it is also an accelerator of the learning and decision-making processes. Contrary to conventional "wisdom" that says Set-Based requires additional work up front but that it will pay off in avoiding later rework, we hope you see that it is in fact going to help you do far less work in the front-end learning (by enabling Decision-Focused Learning) and less work in overbuilding during development (by enabling clear visibility to where the real limits are). So it will reduce your workload, even ignoring the potential reductions in rework. (And that doesn't even mention the future work reductions due to the reusability of that Set-Based knowledge!)

After teaching the material in Part I to an organization, we often ask them "Where in your process should you be establishing 'Success is Assured'?" Typically, the quick answer is before detailed design work begins. But the ensuing discussion usually points to some stage earlier in the preliminary design work.

However, as those teams start to gain experience with Set-Based methods as a learning and decision-making accelerator, they start to see the potential for establishing "Success is Assured" even earlier. And with their eyes opened to the value of that, they start to challenge the critical decisions that they are making earlier without sufficient knowledge.

Notice in Part II that they effectively push toward establishing that "Success is Assured" prior to writing the requirements for the aircraft they want! From the perspective of the development process, they push "Success is Assured" prior to the Request for Proposal (RFP)! We have never dared to tell our clients that it should be done that early, but our clients often come to that conclusion on their own. Where they can have the biggest impact is in shaping the demand coming from their customers, as so many of the critical design decisions are baked into the "requirements."

Where it belongs for your organization depends on your organization and its customers. But it is an important question that you should try to answer aggressively.

A mechanism that allows accelerated collaborative learning and decision-making in such a way that even non-experts can see that "Success is Assured" prior to decision-making should be pretty appealing to most executive leaders.

As a closing thought for you executive leaders wanting to implement that in your own organizations: You need to make sure your organization understands that "True North" is the priority, and that if they are running into obstacles getting in the way of "True North," then they need to make those obstacles visible to you so that you can clear them. And then you need to actually clear them! (See Chapter 6 for more on that.)

Appendix I: Character List Quick Reference

For those who forget who is who as they read Part II, this quick reference should help. Also, note the first names starting with A–J are IRT employees, L–N are Navy officers, P–R are PFC employees, and S–T are UA employees.

Name, Company: Brief Description

Alex Taylor, IRT: a "Success is Assured" Mentor (SAM) skilled at leading learning discussions and utilizing the visual models, particularly the Causal Map; reports to Carl Garcia

Brett Lee, IRT: a SAM who works with Alex, both reporting to Carl

Carl Garcia, IRT: manager of the Process Excellence organization, which includes the group of SAMs; reports to David Zeller

David Zeller, IRT: manager of the Functional Excellence organization; reports to CEO Brenda Caine

Jack Hawkins, IRT: manager of the Program Execution organization; reports to CEO Brenda Caine

Joe Rivera, IRT: one of the top Program Chief Engineers; reports to Christine Dumas, who reports to Jack Hawkins

Laura Ramirez, Navy: acquisitions lead

Maria Green, Navy: special projects systems engineer

Mark Jackson, Navy: surveillance systems expert

Nathan Harris, Navy: mission support systems lead

Paul Lopez, PFC: a SAM who often works with Rick, capturing the discussion in visual models

Rick Moore, PFC: senior customer-facing propulsion engineer

Scott Park, UA: engineer in the Testing and Analysis organization

Teresa Hernandez, UA: senior systems engineer

Tim Lewis, UA: program manager for Naval products

Appendix II: A Brief History of Lean Product Development and Set-Based Concurrent Engineering

This appendix is for those who want to learn more of the background and history of the concepts taught in this book and that fed into Pratt & Whitney, as documented in Chapter 1. Rather than a simple set of endnotes, this appendix will provide a historical timeline to give the big picture of how the conceptual underpinnings of this book developed, along with specific references that you can use to dig further into the conceptual underpinnings.

To keep it a "brief" history, we are focusing specifically on Lean Product Development and Set-Based Concurrent Engineering (SBCE), as distinct from

- Lean Production and the Toyota Production System
- Lean Management and the Toyota Management System
- Adaptations of Lean Production or Lean Management techniques to Product Development
- Project Management and Portfolio Management
- Systems Engineering and Requirements Management
- General Concurrent Engineering
- Lean Six Sigma and Design for Six Sigma
- Theory of Inventive Problem Solving (TRIZ)
- Knowledge Management
- Risk Management
- Phase Gate and related systems

All of these certainly have had significant contributions to the evolution of Lean Product Development and SBCE, but they are adequately documented elsewhere. Including all of those here would make it very difficult to see the specific developments that we want to emphasize.

1901: After traditional development failures, Wilbur Wright tells his brother Orville that "men would not fly for fifty years."[1]

1901: The Wright Brothers identify three Knowledge Gaps and stop designing airplanes until they close them; "Success is Assured" is born.

1902: The Wright Brothers use a wind tunnel to test hundreds of wings, fully characterizing their design space; Set-Based is born.[2,3]

1903: The Wright Brothers establish that "Success is Assured" and soon after succeed in flying the first airplane.[4,5,6,7]

1910–1960s: The aerospace industry continues using such practices.

1940s–today: Toyota leverages such practices (hiring many aerospace engineers after World War II).

1989: Allen Ward publishes his thesis *A Theory of Quantitiative Inference Applied to a Mechanical Design Compiler*, where he reinvents some Set-Based practices to enable a consistently convergent design process, leading to his subsequent search for and discovery of SBCE at Toyota.[8]

1995: Allen Ward, Durward Sobek, et al. publish "The Second Toyota Paradox" in the *Sloan Management Review*, explaining SBCE and other Lean Product Development (LPD) practices to a broader audience.[9]

1997: The National Center for Manufacturing Sciences (NCMS) creates a consortium to find Best Practices, which later evolves into NCMS's Lean Product Development Initiative lead by Mike Gnam, Allen Ward, and Michael Kennedy.

1998: Sobek, Liker, and Ward publish a follow-up paper on Toyota in the *Harvard Business Review*, bringing even broader attention to SBCE.[10]

1999: Sobek, Ward, and Liker publish "Toyota's Principles of Set-Based Concurrent Engineering" in *Sloan Management Review*.[11]

---

[1] M.W. McFarland (editor), *The Papers of Wilbur and Orville Wright*, New York, NY: McGraw-Hill, 1953, p. 934.

[2] M.W. McFarland (editor), *The Papers of Wilbur and Orville Wright*, New York, NY: McGraw-Hill, 1953.

[3] American Institute of Aeronautics and Astronautics (AIAA) Wright Flyer Project. Online: http://www.wrightflyer.org/WindTunnel/testing1.html (accessed September 20, 2012).

[4] J. Tobin, *To Conquer the Air: The Wright Brothers and the Great Race for Flight*, New York, NY: Free Press, 2003, pp. 23 and 192.

[5] P.L. Jakab, *Visions of a Flying Machine: The Wright Brothers and the Process of Invention*, Washington, DC: Smithsonian Institution, 1990.

[6] M. Eppler, *The Wright Way: Seven Problem-Solving Principles from the Wright Brothers that Can Make Your Business Soar.* New York, NY: Anacom, 2004, pp. 19 and 66.

[7] J.D. Anderson, *The Airplane: A History of Its Technology*, Reston, VA, AIAA, 2002, pp. 64–78.

[8] A. Ward and W. Seering, Quantitative inference in a mechanical design compiler, *ASME Journal of Mechanical Design* 115 (1993), 29–35.

[9] A.C. Ward, J.K. Liker, J.J. Cristiano, and D.K. Sobek II, The second Toyota paradox: How delaying decisions can make better cars faster, *Sloan Management Review* 36 (1995), 43–61.

[10] D.K. Sobek II, J.K. Liker, and A.C. Ward, Another look at Toyota's integrated product development, *Harvard Business Review* 76 (July–Aug 1998), 36–49.

[11] D. Sobek II, A.C. Ward, and J.K. Liker, Toyota's principles of Set-Based concurrent engineering, *Sloan Management Review* 40 (1999), 31–40.

2000: NCMS publishes the consortium's findings, highlighting the SBCE paradigm as the key differentiator in "Product Development Process: Methodology and Performance Measures."[12]

2001: The *Manifesto for Agile Software Development* is published, highlighting the key learning from the development of agile software methods during the late 1990s.[13]

2003: The first book on Toyota's Product Development System is published: *Product Development for the Lean Enterprise* by Michael Kennedy (foreword by Allen Ward).[14]

2004: Allen Ward dies in a plane crash — a huge loss to the LPD/SBCE community.

2004: Targeted Convergence Corporation (TCC) is founded; provides training of organizations in Toyota LPD and Set-Based practices.

2006: Morgan and Liker publish the book *The Toyota Product Development System*.[15]

2007: Durward Sobek finishes and publishes Allen Ward's book, *Lean Product and Process Development*.[16]

2008: *Ready, Set, Dominate* is published by Michael Kennedy, Ed Minnock, and Kent Harmon, describing experiences deploying LPD and SBCE at TCC clients, including a case study of Teledyne.[17]

2008: *Understanding A3 Thinking* is published by Durward Sobek and Art Smalley.[18]

2008: *Managing to Learn* is published by John Shook.[19]

2008: The Lean Product and Process Development Exchange (LPPDE) is founded and holds its first conference, bringing together numerous thought leaders in the space.

2009: Singer, Doerry, and Buckley publish "What is Set-Based Design?" in response to Admiral Sullivan's call for improvements to the Naval Sea Systems Command's design tools, which called out the need for Set-Based design. (Originally presented at ASNE Day 2009.)[20]

---

[12] NCMS, Product development process: Methodology and performance measures: Final report, Project no. 130120, Ann Arbor, MI: NCMS, January 31, 2000.

[13] Manifesto for Agile Software Development. Online: http://agilemanifesto.org.

[14] M.N. Kennedy, *Product Development for the Lean Enterprise: Why Toyota's System Is Four Times More Productive and How You Can Implement It*, Richmond, VA: The Oaklea Press, 2003.

[15] J.M. Morgan and J.K. Liker, *The Toyota Product Development System: Integrating People, Process, and Technology*, New York, NY: Productivity Press, 2006.

[16] A. C. Ward, *Lean Product and Process Development*, Cambridge, MA: Lean Enterprise Institute, 2007.

[17] M.N. Kennedy, J.K. Harmon, and E.R. Minnock, *Ready, Set, Dominate: Implement Toyota's Set-Based Learning for Developing Products and Nobody Can Catch You*, Richmond, VA: The Oaklea Press, 2008.

[18] D.K. Sobek II and A. Smalley, *Understanding A3 Thinking: A Critical Component of Toyota's PDCA Management System*, New York, NY: Productivity Press, 2008.

[19] J. Shook, *Managing to Learn*, Cambridge, MA: Lean Enterprise Institute, 2008.

[20] D.J. Singer, N. Doerry, and M.E. Buckley, What is Set-Based design?, *ASNE Naval Engineers Journal* 121, no. 4 (2009), 31–43.

2010: *The Lean Machine* is published by Dantar Oosterwal, documenting the application of LPD at Harley-Davidson, a client of Allen Ward.[21]

2010–2011: The U.S. Navy publishes several papers on their use of Set-Based design on the Ship-to-Shore Connector program, the first U.S. Navy program to apply Set-Based practices.[22]

2011: Brian Gracias of Pratt & Whitney attends LPPDE and contacts Michael Kennedy and Ron Marsiglio, formerly CEO of TCC client Teledyne, and begins working with TCC.

2013: Kennedy, Sobek, and Kennedy publish "Reducing Rework by Applying Set-Based Practices Early in the Systems Engineering Process" in the *Systems Engineering Journal*, describing how SBCE impacts the larger systems engineering processes in companies.[23]

2013: *Knowledge Based Product Development: A Practical Guide* is published by Bob Melvin, documenting the lessons learned from practicing LPD at Teledyne, a client of TCC.[24]

2014: *The Lean Mindset: Ask the Right Questions* is published by Mary and Tom Poppendieck, documenting some of the cross-fertilization between Agile Software Development and LPD, including progress made at Intel, a client of TCC.[25]

2014: Durward Sobek updates Allen Ward's book with some of Ward's later thinking that wasn't in his original manuscript, releasing it as the second edition of *Lean Product and Process Development*.[26]

2017: The American Society of Naval Engineers (ASNE) hosts the first of its Design Sciences Series focused on Set-Based Design, the first conference anywhere specifically on Set-Based, highlighting the several different U.S. Navy projects that applied Set-Based design over the past decade.

---

[21] D.P. Oosterwal, *The Lean Machine: How Harley-Davidson Drove Top-Line Growth and Profitability with Revolutionary Lean Product Development*, New York, NY: AMACOM, 2010.

[22] W. Mebane, C. Carlson, C. Dowd, D. Singer, and M. Buckley, Set-Based design and the Ship to Shore Connector, ASNE Day (February 9–10, 2011), Arlington, VA, 2011.

[23] B.M. Kennedy, D.K. Sobek II, and M.N. Kennedy, Reducing rework by applying Set-Based practices early in the systems engineering process, *Systems Engineering Journal* 17, no. 3 (Autumn 2014), 278–296 (first published online May 21, 2013).

[24] R.G. Melvin II, *Knowledge Based Product Development: A Practical Guide*, self published, 2013.

[25] M. Poppendieck and T. Poppendieck, *The Lean Mindset: Ask the Right Questions*, Upper Saddle River, NJ: Pearson Education, 2014.

[26] A.C. Ward and D.K. Sobek II, *Lean Product and Process Development*, Second edition. Cambridge, MA: Lean Enterprise Institute, 2014.

Appendix III: Causal Mapping for Problem-Solving

When solving a problem, it is often useful to brainstorm contributors to the problem and take some time to collect data (or analyze data you have already collected) to point to the most likely suspects, thereby focusing your Causal Mapping efforts. One benefit is that it focuses the team on *Looking* at the problem, *Asking* questions, and *Modeling* the facts of the situations, before engaging in the Causal Analysis that will often lead to visibility of solutions, and thus *jumping to solutions*.

On the other hand, the Causal Map can be a great place to do that brainstorming; so, with a little discipline, jumping early into Casual Mapping can provide valuable insight, as long as the tendency to jump to solutions is managed (redirected to the Alternatives Evaluation Matrix, as mentioned in Chapter 4).

So, once the team is ready to begin Causal Mapping regarding a Problem Description that they have crafted and gotten consensus on, here are the recommended steps of the process, composed as a series of questions that you can keep asking and answering in the form of shapes in the visual Causal Map.

1. *Ask*: **What is the problem that you are trying to solve?** Add it to the map as a Causal Statement (or perhaps a Fuzzy Statement if it may be several problems), marked with a Problem flag and a Key flag (since it is the focus of this Causal Mapping effort). *Ask*: What? When? Where? Who is the victim? (*Do not ask*: Who is to blame?) Most importantly, *ask five times*: So what? (What is the significance?) That may lead to a better Problem Description to focus on.

2. *Ask*: **Why? What are the likely causes?** *Ask*: **What conditions had to also exist for that to happen?** Add each (there's usually more than one) to the map as Causal Statements. (Use question mark, check, and X flags to document whether that cause is confirmed or denied for the actual cases you are trying to solve.) Consider marking the connectors with "and" or "or" depending on whether all need to occur together to have the effect ("and") or whether any one by itself is sufficient to have the effect ("or").

3. ***Ask*: What are the causes of those causes? (Repeat step 2 for each, five times at least.)** Continue to dig until you get to the past decisions that need to be changed to prevent the chain of causes and that you can potentially change. In other words, continue until you find the limit or limits that some of your past decisions violated, resulting in the failure. For those causal decisions, use Decision shapes instead of general Causal shapes, and ask for the unit of measure and maximum range of values. And for the violated limits, use Relation shapes.

4. ***Ask*: What customer interests may be impacted by changing the decisions identified in step 3?** Add each to the map as a Decision shape (with a Target flag) and document its unit of measure, range of values, and target values. Among those should be one or more customer interests corresponding to the problem identified in step 1.

5. ***Ask*: What are the trade-offs between those decisions we need to change and the various customer interests? *Ask*: As one goes up, how much does the other go up or down?** (The answer will often be, "Well, it depends...." In which case, go to step 6.) Capture each trade-off between Decision shapes as a Relation shape connecting those Decision shapes. If the equation or data for the curve is not known but it is known how to get it, mark it with a "T?" flag; if it is unknown how to get that equation or data, then mark it with an "N?" flag; or if another problem will need to be solved to determine it, mark it with a Problem flag.

6. ***Ask*: It depends on what? What other decisions do those trade-offs depend on? What conditions must exist?** Add each to the map as Decision shapes, documenting the unit of measure and range of values for each. (In brainstorming these, you might use the general Causal or Fuzzy shapes to temporarily capture influencers that need to be broken down into specific Decision shapes.) Return to step 5 whenever you have captured all that a trade-off depends on and complete that Relation shape.

7. ***Ask*: What else might these things depend on? *Ask*: What other customer interests might depend on these things?** (Look for *dangling Decisions* — Decisions that only affect one customer interest — and *Ask*: Why can't I make this arbitrarily high or low to optimize that one customer interest?) Repeat steps 5 through 7 until all trade-offs are identified.

8. **Construct Trade-Off Charts from this map** to show the aggregate trade-offs between the various customer interests, and get customer or marketing input on where they would like your product to be on those trade-off curves.

9. ***Ask*: Are we satisfied with the design window indicated by the trade-off charts that emerged, or is it worth the additional effort to try to move one or more of the limit curves in a specific direction?**

If it is worth the effort (if there is available time), then move to "Causal Mapping for Moving a Limit Curve in a Specific Direction" (Appendix VI) and be prepared to move into blank areas to the side of this map to start new maps as you apply those questions to each limit curve you want to move.

So, starting with step 1 when Causal Mapping the example problem from Chapter 4, they might add this Causal Statement shape to the Map, marked with the Problem flag (because it is a problem) and with the Key flag since it is the

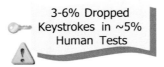

Figure AIII.1   Step 1: The problem we are trying to solve.

focus of this Causal Map (Figure AIII.1).

Repeating steps 2 and 3 may then result in two additional Causal Statements,

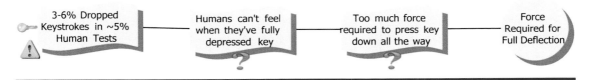

Figure AIII.2   Steps 2 and 3: Answering why down to a key decision.

and then get down to a key Decision (Figure AIII.2).

Continuing with steps 4 and 5 may then result in adding the decision regarding the distance to full deflection of the key, and a relation between that and the force required to depress the key that full deflection (Figure AIII.3).

Figure AIII.3   Steps 4 and 5: Mapping to other decisions that are impacted.

Steps 6 and 7 may initially start bringing in other issues (Figure AIII.4).

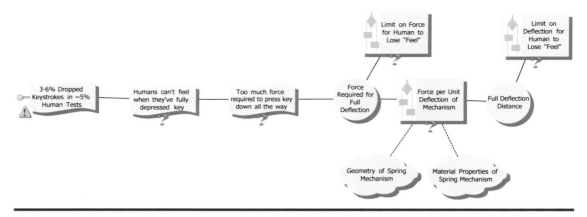

Figure AIII.4   Steps 6 and 7: Mapping to other issues and impacts.

Eventually, you may settle on the Map shown in Figure AIII.5.

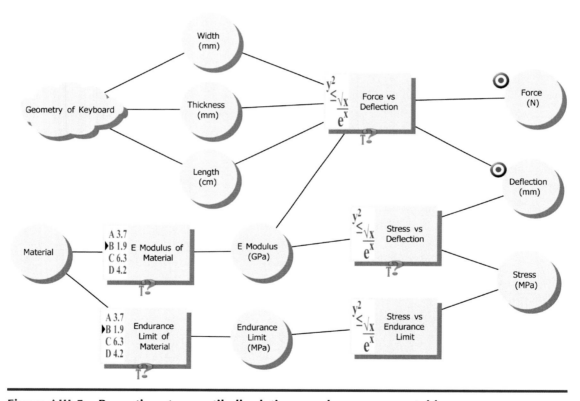

Figure AIII.5   Repeating steps until all relations are known computable.

Step 8 might lead to a chart like the one shown in Chapter 4 (and repeated here in Figure AIII.6).

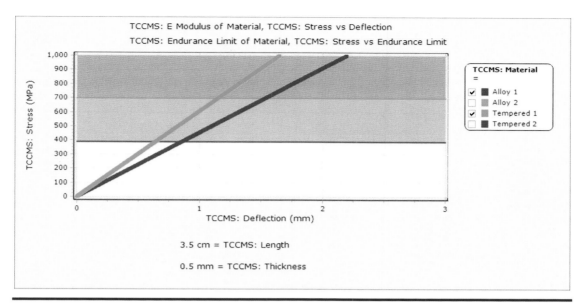

Figure AIII.6   Step 8: Generate Trade-Off Chart from resulting Decision Map.

Appendix IV: Causal Mapping for Making a Set of Decisions

Before the example Problem of Chapter 4 was discovered, the team would have been assigned the task of making those design decisions in the first place. Most of the decisions all depend on the choice of Material, so they may have chosen to make that decision first. Looking just at that decision, they may have created an evaluation matrix with all the alternative materials in the rows and the various properties that might be important in the columns to be evaluated (Figure AIV.1).

From this, they might be torn. Tempered 2 looks best (almost all green), unless it turns out that E-Modulus is most important. In that case, perhaps Alloy 1 is best, but it is pretty bad at the other three. So, is there an alternative for which "Success is Assured"? Maybe all, maybe none; it is impossible to tell without knowing what all that decision may impact and knowing the degree of impact.

So, with "True North" as your guide, Causal Mapping is in order before making this Decision. The Causal Mapping steps to "Make a Set of Decisions" are

1. *Ask*: **What are the decisions that you are trying to make?** Add each to the map as Decision shapes, and document the unit of measure and range of values of each.
2. *Ask*: **What are the customer interests that you might impact (for better or worse)?** (If there are none, then you can choose anything for those decisions in the map. *Ask*: Why can't we make each of these decisions as high (or low) as we want?) Add each customer interest to the map as a

Material	Row Status and Reason	E-Modulus 1000000 psi	Yield 1000 psi	UTS 1000 psi	Endurance Limit 1000 psi
Alloy 1	Best if E Mod is key	27	120	160	68
Alloy 2		32	130	175	38
Tempered 1		34	220	290	94
Tempered 2	Near-best in 3 of 4	34	310	340	88

Figure AIV.1 **Alternatives Evaluation Matrix for different materials and their properties.**

Decision shape (with Target flag), and document the unit of measure, range of values, and target values of each.

3. *Ask*: **What is the trade-off between those?** *Ask*: **As one goes up, how much does the other go up or down?** (The answer will often be, "Well, it depends…" In which case go to Step 4.) Capture each trade-off between Decision shapes as a Relation shape connecting those Decision shapes. If the equation or data for the curve is not known, but it is known how to get it, mark it with a "T?" flag; if it is unknown how to get that equation or data, then mark it with an "N?" flag; or if another Problem will need to be solved to determine that, mark it with a Problem flag.

4. *Ask*: **It depends on what? What other decisions do those trade-offs depend on?** What conditions must exist? Add each to the map as Decision shapes, documenting the unit of measure and range of values for each. (In brainstorming these, you might use the general Causal or Fuzzy shapes to temporarily capture influencers that need to be broken down into specific Decision shapes.) Return to step 3 whenever you have captured all that a trade-off depends on, and complete that Relation shape.

5. *Ask*: **What else might these things depend on?** *Ask*: **What other customer interests might depend on these things?** (Look for *dangling Decisions* — Decisions that only affect one Customer Interest. *Ask*: Why can't I make this arbitrarily high or low to optimize that one Customer Interest?) Repeat steps 3 through 5 until all trade-offs are identified.

6. **Construct Trade-Off Charts from this map** to show the aggregate trade-offs between the various Customer Interests, and get customer or marketing input on where they would like your product to be on those trade-off curves.

7. *Ask*: **Are we satisfied with the design window indicated by the Trade-Off Charts that emerged, or is it worth the additional effort to try to move one or more of the limit curves in a specific direction?** If it is worth the effort (if there is available time), then move to "Causal Mapping for Moving a Limit Curve in a Specific Direction" (Appendix VI) and be prepared to move into blank areas to the side of this map to start new maps as you apply those questions to each limit curve you want to move.

So, starting with Step 1, they add a Decision shape for the Material selection decision and mark it with a Key flag since that is the focus of this Causal Mapping effort (Figure AIV.2).

Figure AIV.2   Step 1: the decision we need to make (what material to use).

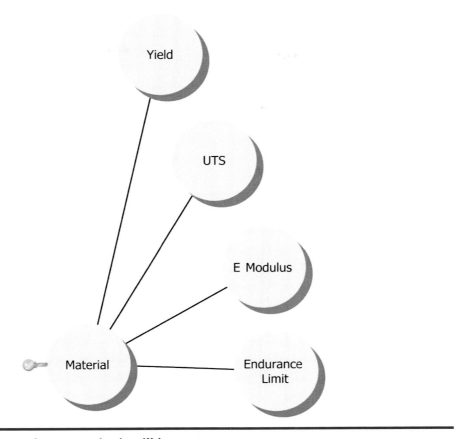

Figure AIV.3   Step 2: the properties it will impact.

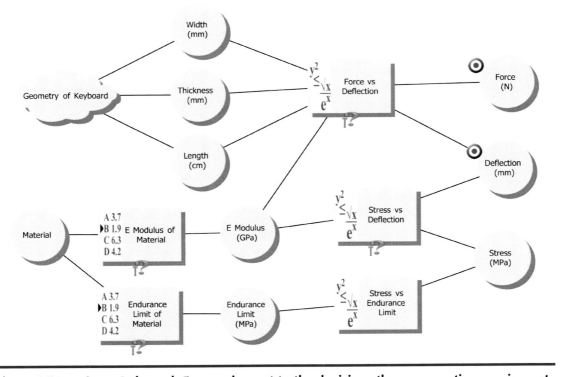

Figure AIV.4   Steps 3 through 5: mapping out to the decisions those properties may impact.

In the preceding Alternatives Evaluation Matrix, they have already identified four things it will impact. You might argue that those are not customer interests, and you would be correct. So, effectively, we jumped to step 5. That is okay, as we will repeat steps 3 through 5 repeatedly anyway in order uncover all the rest of the customer interests. (Figure AIV.3).

As they continue to ask what those things might impact, they will tend to build out the larger Causal Maps we saw earlier, perhaps at first ignoring Cost (Figure AIV.4).

Appendix V: Causal Mapping for Achieving or Improving a Customer Interest

When trying to achieve, improve, or optimize a particular customer interest (e.g., reduce cost, reduce weight, increase battery life, reduce noise, etc.), the suggested steps are identical to the previous set, except for the first two:

1. *Ask*: **What is the customer interest that you are working to improve?** Add it to the map as a Decision shape (with Target and Key flags), and document its unit of measure, range of values, and target values.
2. *Ask*: **What are the competing customer interests that might be made worse?** (If there are none, then improving the decision identified in Step 1 will be trivial. *Ask*: Why can't we make the decision identified in Step 1 as high (or low) as we want?) Add each to the map as Decision shapes (with target flag), and document the unit of measure of each.

Steps 3 through 7 are the same as steps 3 through 7 of "Causal Mapping for Making a Set of Decisions."

Given the bulk of the steps are identical, mapping the previous Keyboard Material Selection example to this case, starting with a focus on reducing the 'Force' Customer Interest, should be easy. We will not bother with that here.

Further, Part II provided an extensive example of the Navy working with an Infrared Technology supplier, an Unmanned Aircraft supplier, and a Jet Engine supplier to achieve a set of Customer Interests, following these seven steps repeatedly, fleshing out a Causal Map for a full flight mission.

Appendix VI: Causal Mapping for Moving a Limit Curve in a Specific Direction

The last step of each of the preceding three sets of steps (Appendices III through V) pointed to this one. This one is also useful stand-alone. And unlike the previous three, which were very similar to each other after the first few steps, this one is very different from the rest.

This set of steps borrows heavily from the TRIZ (pronounced "trees"; a Russian acronym for "Theory of Inventive Problem-Solving") literature, with a few key differences. First, the TRIZ literature focuses on identifying two-way *contradictions* and then innovating ways to eliminate those. We call those contradictions *trade-offs* because in most product development efforts, the majority of such contradictions will not be eliminated, leaving you with the need to make the right trade-off decisions. Second, most of the trade-offs/contradictions that we wrestle with are not two-way but rather multi-dimensional. Dealing with those just two at a time may be sensible when trying to eliminate them but problematic when needing to make trade-off decisions.

Fortunately, many of the TRIZ methods can be easily adjusted such that they are effective at helping you find ways to move the trade-off curves, even if you cannot eliminate them. (But if you *can* find ways to eliminate the need to make those trade-offs, all the better!) We employ variants of the TRIZ methods here as we have with the rest of these Causal Mapping methodologies, by formulating them as a series of questions that drive knowledge (and identify knowledge gaps) regarding those innovations into the Causal Map that is depicting the trade-offs.

The suggested steps are as follows:

1. *Ask*: **What is the relation defining the limit curve that you are wanting to move?** Add it to the map as a relation shape (with a Key flag) and its related Decision shapes. (Copy that portion of your existing Decision Map.) Make a note in its abstract of what direction you need to move it. What would be ideal?

2. ***Ask*: Is there anything else not yet mapped that this curve might be dependent on?** What conditions must exist? (E.g., if you changed the material used by one of the parts involved, could those new materials change the curve?) If so, add those potential dependencies as additional Decision shapes hanging off that Relation shape, with a "?" flag to indicate its effect needs to be tested. *Ask:* What if there were super-strong, super-fast, super-tiny people helping out, what could they do to move this limit curve? Does that inspire ideas on what else could be changed (and therefore are additional decisions that would affect the limit curve)?

3. ***Ask*: For each of the decisions that the relation is dependent on, are there additional values or options that have not yet been considered that may change the shape of the curve?** (E.g., you may already have a material selection decision but have only considered two types of steel and aluminum. Should you add titanium, carbon fiber, and/or plastics to the options for that decision?) Set the Proposal (light bulb) flag on such Decision shapes to indicate that new options or values have been added to its range. For each Decision shape, *ask*: What if the value was enormous? What if the value was tiny or zero?

4. ***Ask*: Can we move the analysis to the next lower level of detail?** (E.g., if material selection is one of the Decisions, can we replace that decision with the various material properties, and change the relation to compute the curve based on those properties?) Doing so will often result in an entirely different Causal Map and often expose new dependencies that can be exploited. Move to a blank area of the map and start mapping at that lower level.

5. ***Ask*: Can we move the analysis to the "bigger picture"?** Why do the customers want these customer interests? Why does the business want these business interests? What are they each really trying to achieve? Is there another way to satisfy them that avoids this limit curve altogether? Or perhaps that changes the sensitivity to this limit curve? Move to a blank area of the Map and start mapping from those "bigger" interests to see if alternate connections emerge.

6. ***Ask*: Can we move the analysis to an earlier point in time?** Perhaps the solution is to prevent the situation or to pre-condition things to better handle the situation, and so on. Often, the trade-off curves for prevention or avoidance mechanisms are very different than the trade-off curves for resistance or tolerance mechanisms. Move to a blank area of the map and start Causal Mapping with the customer interests of avoiding, preventing, or pre-conditioning to achieve the original customer interest(s).

7. ***Ask*: Can we combine some of the ideas in step 6 with the ideas of steps 4 or 5?** For example, look at the bigger picture at an earlier point in time, or look at the micro-level at an earlier point in time. (For those familiar with Nine Windows or Nine Screens from the TRIZ literature, that may be an appropriate visual model to employ here.)

8. **Based on how promising or risky each alternative map is, make plans to flesh out each of the maps as necessary to make them computable so that they can be incorporated into your design decisions.** And then prioritize based on the cost of those plans and the risk/reward. (Making them computable may involve testing or may involve working with marketing to study the customers or talking with the business leaders.)

To show the real-world applications of each of the preceding steps would require numerous separate examples, as the different techniques are such that when the others won't work, one of them will. Presenting numerous examples to illustrate each of these methods is beyond the scope of this book, primarily because it is already well covered in the existing TRIZ literature. If you want to learn more about the specific techniques embedded in these eight steps, we suggest you explore the following references.[1,2,3,4,5,6,7,8]

[1] G. Altshuller, *The Innovation Algorithm: TRIZ, Systematic Innovation and Technical Creativity*, Worcester, MA, Technical Innovation Center, 1999.

[2] K. Rantanen and E. Domb, *Simplified TRIZ: New Problem Solving Applications for Engineers and Manufacturing Professionals*, Boca Raton, FL, Auerbach, 2008.

[3] V. Fey and E. Rivin, *Innovation on Demand: New Product Development Using TRIZ*, Cambridge, UK: Cambridge University Press, 2005.

[4] D. Gray, S. Brown, and J. Macanufo, *Gamestorming: A Playbook for Innovators, Rulebreakers, and Changemakers*, Sebastopol, CA: O'Reilly Media, 2010.

[5] C.M. Christensen and M.E. Raynor, *The Innovator's Solution: Creating and Sustaining Successful Growth*, Boston, MA, Harvard Business Review Press, 2013.

[6] M. Michalko, *Cracking Creativity: The Secrets of Creative Genius*, New York, NY: Ten Speed Press, 1998.

[7] M. Michalko, *Thinkertoys: A Handbook of Creative-Thinking Techniques*, New York, NY: Ten Speed Press, 2006.

[8] A. Ulwick, *What Customers Want: Using Outcome-Driven Innovation to Create Breakthrough Products and Services*, New York, NY: McGraw-Hill, 2005.

Appendix VII: Problem K-Brief

The example Problem K-Brief shown in Figures AVII.1 and AVII.2 was developed in Chapter 4. (See Chapter 4 for a full discussion of that example.)

The Problem K-Brief is designed to pull a problem-solving process. It begins with a Problem Description that lays out the key objectives and clearly defines when the problem is solved. It then digs to root causes and back out to potential impacts of changing those root causes, and ultimately develops the knowledge required to make the right decisions (the optimal trade-offs) when changing the decisions (the root causes) in order to best solve the identified problem.

A Recommended Storyline

We recommend the Problem K-Brief pull the following story:

- What is the issue that needs to be resolved, the standard not being met, the situation of concern, or the knowledge gap that needs to be closed? What is the impact of that? Who is it impacting?
- What are the objectives of this Problem K-Brief? How do we know the problem is adequately solved? By when does it need to be adequately solved?
- What is potentially contributing to the problem? Are those contributors adequate in frequency and severity to explain the frequency and severity of the problem?
- What are the root causes that we can control (directly or indirectly)? How are those decisions causally related to the objectives we are trying to achieve? What decisions will impact them? What decisions will they impact? What fundamental limits must be respected or worked around? What trade-offs must be made? (A Causal Map is typically the best way to answer these questions.)
- What in the existing situation or environment could be potentially leveraged to help resolve the problem? Are there alternatives in the surrounding environment? Are there alternatives at the subsystem or component level? Are there preventative alternatives? Are there alternatives to reduce or eliminate the impacts of the problem, rather than eliminating the problem?

ℰ PROBLEM | TCCMS: 7 Keyboards are passing automated testing b

PROBLEM

TCCMS: 7 Keyboards are passing automated testing but human testers complain they drop keystrokes

LEAD	ID
TCC Admin	1049960
LAST EDITED ON	STATUS
2017-03-24 13:03	Draft
CATEGORIES	
Material Selection Example (TCCMS)	

ABSTRACT

This is an example Problem... it is intended to show Causal Analysis brainstorming potential causes... testing confirming or denying those... LAMDA Discussion helping to brainstorm more... and ultimately digging to root causes (decisions we control).

Problem Description

Our new keyboard was passing all of our automated mechanical testing. 0% dropped keystrokes; 0% sticky key rate; no mechanical failures. So, we began some final human testing... roughly 5% of the human testers reported 3-6% dropped keystrokes. What is different about those human testers vs. our automated mechanical testing? And how do we fix the keyboard prior to our scheduled release end of this month?

The objective is to revise the automated testing to detect the dropped keystrokes, and then modify the keyboard design to not exhibit those dropped keystrokes. Both prior to the end of this month, so as to not delay product release. And without adding more than $1 in part costs.

Causal Analysis

It may be useful to look at potential differences between the human testing and the automated testing.

It may be useful to look at the sequence of events and brainstorm failure modes. (Some of the failure modes here helped us identify potential differences in the tests listed above.)

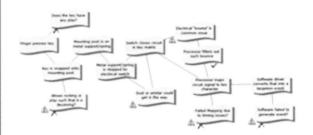

In the weekly cross-team review, Lucy asked if we've seen complaints on the keyboards having poor feel or being tiring to type on. We had seen 2% of such complaints... not enough to warrant immediate action. A few years back, they had a keyboard with such complaints and it turned out that if the force required to press the key all the way down was too high, then people would lose the "feel" of whether or not they pressed the key all the way down. She suggested that such may be our problem here... they felt they pressed it all the way down, but hadn't really.

So, following that line of thinking towards root cause, this initial Causal Map emerged. We plan to continue the Causal Mapping effort from here.

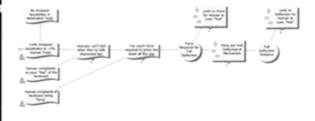

A Simplified Model

For this initial design study, a simplified model of the system was chosen (a rectangular beam supported on both ends with force applied to single point in center) to explore the trade-offs between the material selected and the other design decisions, including:

- d = Deflection of the material due to the force (in)
- F = Force applied to the center of the beam (lbf)
- L = Length of the beam (in)
- W = Width of the beam (in)
- T = Thickness of the beam (in)

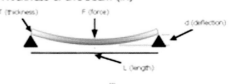

Figure AVII.1   Problem K-Brief tells the problem-solving story with visual models (first half).

but human testers complain they drop keystrokes | Lead: TCC Admin

7) What else might these things depend upon? What other customer interests might depend upon these things?

Notice the Material is dangling... we can solve this Problem by picking some material with super low E Modulus... so, why can't we?

Changing the spring material to get more Deflection for less Force may result in a more flimsy material that wears out more quickly. Ideally, the Stress due to Deflection should never exceed the Endurance Limit of the material.

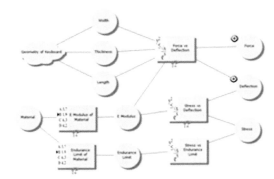

8) Construct the Decision and Relation K-Briefs (construct a Decision Map) so that you can view Trade-Off Charts

Scott was able to collect data on the spring material used, and some alternatives. He also found the following two equations which Jane was able to verify accurately model the existing part through physical testing.

Force vs Deflection

This formula was pulled from XYZ by Joe Johnson:

$$F = \frac{4\,E \cdot W \cdot T^3}{L^3} d$$

Maximum Stress vs Displacement

This formula was pulled off ABC textbook:

$$S = \frac{6\,E \cdot T}{L^2} d$$

Trade-Off Analysis: Force vs Deflection

Alloy 1 offers a better Force vs. Deflection curve than the rest; but it cannot withstand as large of Deflection as the Tempered Steel 1 we are currently using.

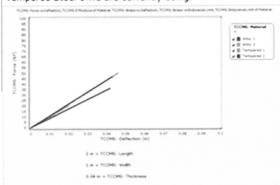

Trade-Off Analysis: Stress vs Deflection

Rearranging the Chart to show the Stress vs. Deflection allows you to see why... the much lower Endurance Limit of Alloy 1. However, as long as we don't need that much Deflection for adequate "Feel" of travel, then Alloy 1 would be the preferred choice.

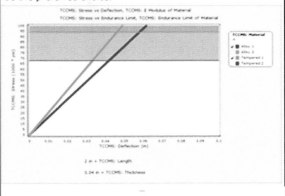

Plan to Close Knowledge Gaps

* To Do once K-Brief's Status >= Discussion

Done	When	Who	What
☑	2017-02-21	Scott	Investigate environmental conditions (go look)
☑	2017-	All	Go look at disassembled keyboard (CMap 2)
☑	2017-02-22	Jeff	Work-up a design for adjusting angle of process of automated tools
☑	2017-02-22	Jane	Re-run automated tests typing same as humans were typing during failures
☑	2017-02-23	Jane	Adjust automated tests by measuring to keyboard
☑	2017-	Jerry	Randomize timing of automated tests
☑	2017-02-24	Jane	Re-run tests with environment adjusted per Scott's findings.
☑	2017-02-28	All	Continue Force vs. Deflection vs. "Feel" Causal Mapping
☑	2017-	Scott	Get the equations and material data.
☑	2017-	Jane	Test the material against the equations.
☐	2017-03-10	Jane	Test a few keyboards modified with Alloy 1 spring.
☐	2017-	Scott	Setup human testing of the modified keyboards to test "Feel" issues.

Trade-Off Analysis: Titanium? Others?

Titanium offers a much better Force vs. Deflection curve than the rest, and its Endurance Limit is still high enough that it offers better Deflection than most. So, if Alloy 1 "Feel" turns out to be inadequate per the testing, we could consider Titanium (or other more exotic materials).

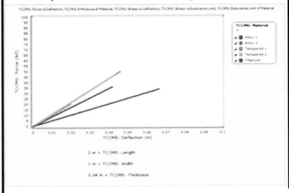

Figure AVII.2   Problem K-Brief tells the problem-solving story with visual models (second half).

■ What are the alternative remedies to the problem, and how do we evaluate them? What are the key targets we should evaluate them against? What are the key trade-offs that will need to be made?

■ What do we need to know to optimize the trade-offs when making those decisions? How do we most quickly close those knowledge gaps? By when does who need to do what to close those gaps?

■ What is the recommended solution? By when does who need to do what in order to put the recommended remedy or remedies into action?

■ What was the result of putting the recommendations into action? Were the results as expected? If not, what further actions are needed? If so, what lessons learned need to be incorporated into our best practices such that we don't have that problem in the future?

Together, those questions pull a quality problem-solving process, and better yet pull a Plan-Do-Check-Act (PDCA)-style continuous improvement process.

Index

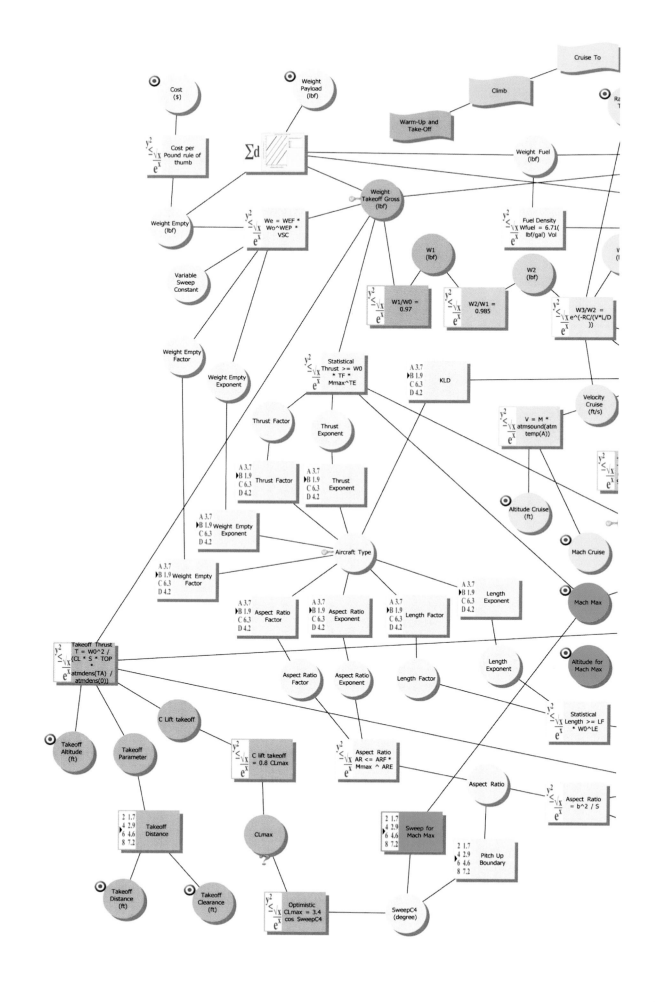

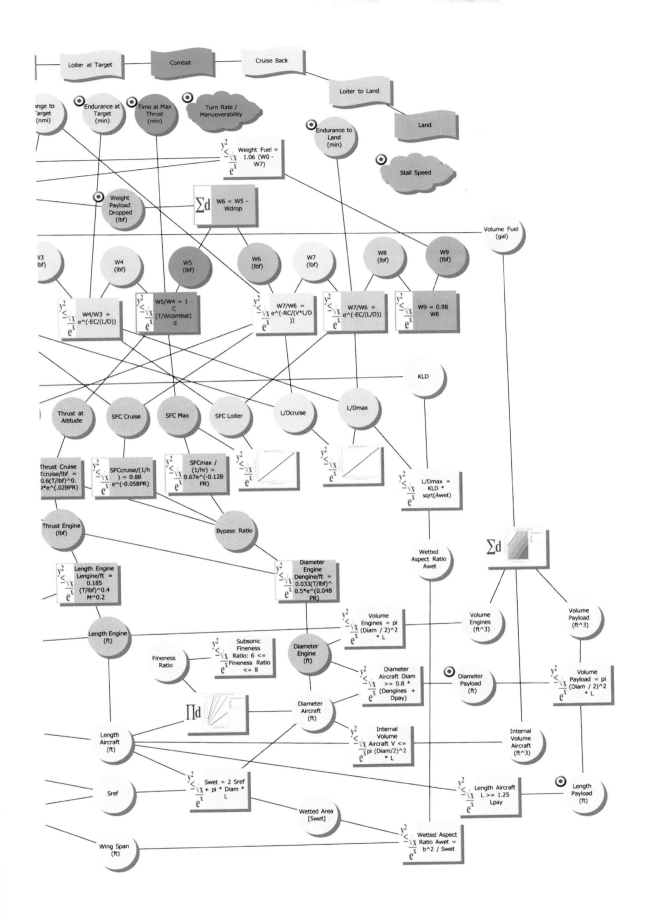

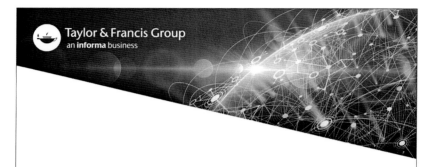